P9-EDJ-548

Proceedings in Life Sciences

Ecological Genetics: The Interface

Edited by
Peter F. Brussard

With Contributions by
R.W. Allard H.L. Carson B. Clarke
M.T. Clegg R.E. Ferrell R.G. Harrison
W.B. Heed A.L. Kahler R.K. Koehn
D.A. Levin S.A. Levin R.C. Richmond
F. Rothhammer J. Roughgarden W.J. Schull
R.R. Sokal

With 39 figures

Springer-Verlag
New York Heidelberg Berlin

PETER F. BRUSSARD
Division of Biological Sciences
Cornell University
Ithaca, New York 14853
USA

Library of Congress Cataloging in Publication Data. Main entry under title: Ecological genetics. (Proceedings in life sciences) Proceedings of a symposium held at Ithaca College, Ithaca, N.Y. June 12-15, 1977 and sponsored by the Society for the Study of Evolution. Includes index. 1. Ecological genetics–Congresses. I. Brussard, Peter F. II. Society for the Study of Evolution. [DNLM: 1. Genetics, Population–Congresses. 2. Ecology–Congresses. 3. Evolution–Congresses. QH455 S678e 1977]
QH456.E26 575.1 78-27196

Printed in the United States of America.

9 8 7 6 5 4 3 2 1

ISBN 0-387-90378-x Springer-Verlag New York Heidelberg Berlin
ISBN 3-540-90378-x Springer-Verlag Berlin Heidelberg New York

Preface

Traditionally, studies in ecological genetics have involved both field observations and laboratory genetic analyses. Comparisons and correlations between these two kinds of data have provided valuable information on the genetic strategies behind the evolutionary adaptations of species and their component local populations. Indeed, much of our current understanding of the dynamics of evolutionary processes has come from syntheses of ecological and genetic information. Since the recent discovery of abundant markers in the form of protein polymorphisms, scientific interest in the connections between genetics and ecology has quickened considerably.

This volume contains the proceedings of the Society for the Study of Evolution's symposium, Genetics and Ecology: The Interface, held at Ithaca College, Ithaca, New York, June 12-15, 1977. This particular topic was selected because of a general feeling that a significant integration of genetics and ecology has developed in the last decade or so. Most ecologists no longer believe that each species has a characteristic and constant birth, death, and development rate, habitat preference, and so on, but that these parameters vary among populations and are at least partially under genetic control and subject to natural selection. Similarly, few population geneticists still view any species as infinitely large, panmictic, constant in numbers, and distributed evenly throughout its range. It is clear that ecology and genetics must interface when one asks such questions as, how does the genotypic composition of a population influence its numbers? How does the abundance of organisms influence a population's genetic composition? Given an observed set of habitat and niche parameters, how have species evolved their present adaptive characteristics? Or, what are the relative roles of genetic and ecological factors in the speciation process?

Despite the recognition that such an interface exists, a genuinely synthetic science of population biology has yet to be forged from the individual disciplines of population ecology and population genetics. Ecology still deals primarily with the numbers of organisms and the factors determining these numbers; and few ecologists collect data which are of specific interest to geneticists. Likewise, population geneticists, currently preoccupied with explaining the selective significance, or nonsignificance, or protein polymorphisms, rarely incorporate much relevant ecological data--or theory--into their interpretations of observed patterns.

These meetings were designed to encourage the dialogue between ecologists and geneticists and perhaps to promote collaborative investigations. The organizing committee, consisting of Ross MacIntyre, James Fogleman, and myself, tried to select symposium speakers who are attempting to integrate ecological and genetic parameters in their research. At the same time we tried to achieve a reasonable balance between theory and experiment, between plants and animals, and between *Drosophila* and other organisms. In selecting these 12 speakers, we doubtless committed a few serious sins of omission. However, I do feel that we have achieved a broad-based and topically balanced symposium, and that this book will be of considerable interest to ecologists and geneticists alike.

In addition to the organizing committee, many people helped me in numerous ways during the preparation of this volume. Simon Levin, Richard Harrison, and Alan Hastings contributed many valuable suggestions during the planning stage. Karen Kelsey and Mindy McCauley typed the final manuscript, and Monica Howland redrew several of the figures. Trudy Brussard and Mindy McCauley assisted with various editorial chores, and Steven Gourley prepared the subject index. I am grateful to them all.

Ithaca, New York
October 1978

Peter F. Brussard

List of Contributors

Allard, R.W., Department of Genetics, University of California, Davis, California 95616

Carson, H.L., Department of Genetics, University of Hawaii, Honolulu, Hawaii 96822

Clarke, B., Genetics Research Unit, University Hospital, Clifton Boulevard, Nottingham N672UH, England

Clegg, M.T., Department of Botany, University of Georgia, Athens, Georgia 30602

Ferrell, R.E., Center for Demographic and Population Genetics, University of Texas Health Science Center, P.O. Box 20334, Houston, Texas 77025

Harrison, R.G., Department of Biology, Yale University, New Haven, Connecticut 06520

Heed, W.B., Department of Ecology and Evolutionary Biology, University of Arizona, Tucson, Arizona 85721

Kahler, A.L., Department of Genetics, University of California, Davis, California 95616

Koehn, R.K., Department of Ecology and Evolution, State University of New York, Stony Brook, New York 11794

Levin, D.A., Department of Botany, University of Texas, Austin, Texas 78712

Levin, S.A., Section of Ecology and Systematics, Cornell University, Ithaca, New York 14853

Richmond, Rollin C., Department of Zoology, Indiana University, Bloomington, Indiana 47401

Rothhammer, F., Department of Cellular Biology and Genetics, University of Chile, Santiago

Roughgarden, J., Department of Biological Sciences, Stanford University, Stanford, California 94305

Schull, W.J., Center for Demographic and Population Genetics, University of Texas Health Science Center, P.O. Box 20334, Houston, Texas 77025

Sokal, R.R., Department of Ecology and Evolution, State University of New York, Stony Brook, New York 11794

Contents

5. Plants

6. Coda

1 Theory

On the Evolution of Ecological Parameters

S.A. Levin

A. Introduction

The wedding (or as W. Provine has described it, the cohabitation) of ecology and population genetics requires an integration of modes of understanding developed at very different levels of organization. Evolution may be quite appropriately viewed either in terms of changes in the phenotypic population, or of changes in frequencies of genetic units. Quite obviously, however, different questions assume importance from those different aspects. One whose primary interests lie at either of those levels may recognize that events at the other have a significant influence on processes of interest, but the detailed mechanics may seem as irrelevant as does a wiring diagram to one whose goal is to operate a television set.

When interest centers upon how organisms adapt to their environments, it is obvious that genetic events such as mutation, segregation, and recombination along with the effects of dominance, linkage, and epistasis define the particular selection algorithm; but it is also true that that algorithm may be only one of many possible ones which lead to the same adaptive end point. A similar distinction of algorithm and task exists in theories of brain function (Marr 1978), where the details of a pattern recognition algorithm are relevant to totally different issues than are questions of function.

The principal difficulty with the "adaptive" approach is the identification of the end result, and the goal of much recent evolutionary theory has been to determine what in fact is the "purpose" of the genetic algorithm. This teleological point of view is born of an almost religious acceptance of the principle that the algorithm may be thought of as leading somewhere; but of course this point of view is fraught with potential pitfalls (see Lewontin 1977, 1978). Gould (1977) points out that Darwin's own view of his theory of "descent with modification" was antipodal to this usage: "Darwin argues that evolution has no purpose. Individuals struggle to increase the representation of their genes in future generations, and that is all (Gould 1977)."

The other approach derives from Fisher (1930), who introduced almost in passing the problem of "the deterioration of the environment." To many population geneticists, nursed in this tradition, the interface of ecology and genetics has meant the introduction of ecological considerations as only slight perturbations upon the classical models and paradigms of population genetics. This too falls far short of providing the total view of the evolutionary process. That total view will necessarily be a novel one, blending both approaches. The synthesis does not yet exist; this paper is an attempt to review the various approaches and provide an evaluation of the state of the art.

B. Theoretical mechanics

The starting point for this survey is an approach which derives from the classical mathematical theory of population genetics. This has an intrinsic appeal for the mathematician because, right or wrong, it is at least axiomatic. It must be borne in mind that it is also concerned with the details of an algorithm, one which we know to be incorrect since it is necessarily oversimplified. The hope must be that qualitative conclusions may be derived which remain inviolate as additional details are appended to the model, and that these conclusions will hence emerge as the fundamental truths of evolution by natural selection. In the search for the appropriate conclusions, the evolutionary point of view may be helpful in providing direction, but it may also be misleading because our evolutionary intuition is often stronger than it has a right to be. We return to this problem in later sections.

The framework which I outline here is developed in the two-species case in Levin and Udovic (1977); but it is a natural extension of the approach of Wright (1955) and more recently Roughgarden (1971), Charlesworth (1971), and Levin (1972). Numerous authors have utilized frequency-dependent selection models to examine intergenotypic competition (see for example Wright 1969, Schutz and Usanis 1969), and others (Norton 1928, Charlesworth and Giesel 1972, Demetrius 1975) have studies selection in age-structured populations. Recently much attention has been paid to models incorporating population density in approaches to density-dependent selection (MacArthur 1962, Anderson 1971, Roughgarden 1971, Charlesworth 1971, and Clarke 1972); and models of interacting populations using this format have been considered by Mode (1958), Jayakar (1970), Yu (1972), Levin (1972), Stewart (1972), Leon (1974), Roughgarden (1976, 1977), Ginzburg (1977b), Leon and Charlesworth (1977) and others.

Consider m populations, generally of different species X^k, $k=1,\ldots,m$ with densities N^k and each segregating at a single autosomal locus with alleles A_j^k, $j=1,\ldots,n_k$. Age structure, size structure, and spatial distribution are ignored, as are interactions with other loci (although the approach may be extended). Mating in each population is random, generations are discrete (the continuous version can be introduced; but it is an approximation), population size is large enough so that stochastic effects may be ignored. A further assumption is that the generation time for each population is the same, although this can be relaxed (Levin and Udovic 1977).

Define w_{ij}^k to be the expectation of offspring (of all possible genotypes) of zygotes carrying alleles A_i^k and A_j^k (Charlesworth 1971). w_{ij}^k will be allowed to vary from generation to generation as a function of population densities and allelic frequencies. Let p_i^k be the frequency of A_i^k in the current zygotic population. Then the overall dynamics of the system are described by a set of $\sum_{k=1}^{m}(n_k + 1)$ difference equations for the values (denoted by primes) of p_1^k and N^k after one time interval,

$$(p_i^k)' = p_i^k w_{i\cdot}^k / \bar{w}^k; \quad (N^k)' = N^k \bar{w}^k, \quad k=1,\ldots,m; \; i=1,\ldots,n_k. \qquad (1)$$

Here $w_{i\cdot}^k = \sum_j p_j^k w_{ij}^k$ is the mean fitness associated with the ith allele, and $\bar{w}^k = \sum_i p_i^k w_{i\cdot}^k$ represents the mean fitness for the population X^k. Clearly $\sum_i p_i^k = 1$ for each k, since the p_i^k represent gene frequencies; and this in turn implies (from (1)) that $\sum_i (p_i^k)' = 1$, which is essential for consistency. Thus only $\sum_k n_k$ of the equations (1) are independent. I assume $w_{ij}^k = w_{ij}^k(\underset{\sim}{p}^1, \underset{\sim}{p}^2, \ldots, \underset{\sim}{p}^m, N^1, \ldots, N^m)$ where $\underset{\sim}{p}^k$ is the vector of allele frequencies in population k and where the functions w_{ij}^k are continuously differentiable on the *feasible region* defined by $p_i^k \geq 0$, $N^k \geq 0$ for all i,k.

The framework (1) is sufficiently general, to encompass problems in density-dependent selection, genetic feedback, and coevolution. It can be modified to include multiple loci (see for example Pimentel, Levin, and Soans 1975) or stochastic effects; yet it contains very little in the way of ultimate truths about natural selection. What one usually does as a first step is to study the properties of (1) as a dynamical system, examining for the equilibria which are defined by the system

$$w_{i\cdot}^k(\underset{\sim}{p}^1, \ldots, \underset{\sim}{p}^m, N^1, \ldots, N^m) = 1, \quad k=1,\ldots,m; \; i=1,\ldots,n^k$$

$$\sum_i p_i^k = 1, \qquad (2)$$

and for boundary equilibria, and then determining the stability properties of those equilibria. A more complete analysis would examine (1) for periodic solutions such as those suggested by the work of Krebs et al. (1969) and for more general solutions and phenomena (Oster 1975; Asmussen and Feldman 1977), but virtually nothing has been done in this vein.

Note that the system (2) is comprised of $\sum_{k=1}^{m} (n_k + 1)$ equations in the same number of variables. Those equations are, however, nonlinear, so that it is not automatic that there are any solutions, and it will often be the case that there are multiple solutions. Further, only solutions for which all values are positive are of interest, and this places an additional restriction.

Fairly complete dynamic analyses of special cases of (1) are provided in Levin and Udovic (1977), Leon (1974), Roughgarden (1976, 1977), and Ginzburg (1977a,b), but these will not be repeated here. With all deference to those authors, the analyses are straightforward exercises in elementary mathematics. Other problems treated in those papers are of more interest, however, and we shall return to them following the next section.

The strength of the system (1) is in its simplicity. If for example a stable polymorphic equilibrium exists, it is fairly easy to interpret why it does from the few parameters in the equations. When equations begin to become complicated with recombination fractions and epistatic relationships, interpretation falls within the province of the divine and drives most of us, in one way or another, to aspire to the next level of organization. Further, as has been often pointed out, an elephantine model with many parameters can be easily made to wag any of several tails and trunks, and more specifically can be made to exhibit virtually any dynamical behavior possible (see for example Smale 1976).

However (1) rests on a minimum of hypotheses and thus provides a good basis for feasibility arguments. For example, Ford (1964) and Pimentel (1961, 1968) have suggested that a selective advantage for the rare type is a sufficient mechanism to account for the coexistence of competitors. The equations (1) allow demonstration that this is at least theoretically correct, and also permit delineation of the ranges of the selection parameters which will permit such coexistence. Similar arguments have been applied to the balance between selection for resistance to predation and selection for rapid growth in the absence of predation. This was suggested by Pimentel (1961, 1968) as having fundamental implications for biological control, and has been analyzed by the above techniques by Levin (1972). Among other things, it may be shown that under Pimentel's assumptions the hypothesized balance requires the maintenance of a polymorphism by overdominance; and this overdominance is characteristic of perhaps only a small range of gene frequencies and population densities. MacArthur (1962), Roughgarden (1971), Charlesworth (1971), Clarke (1972), and Anderson (1971) consider particular applications to the problem of r versus K selection and with similar results.

C. The Macroscopic Approach

Natural selection deals with the differential survival and fecundity of individuals; a fundamental problem of population genetics is the determination of the implications for the population as a whole. Although such an extrapolation must be logically consistent, it is far from the tautology which some critics (Peters 1977) have mistakenly assumed. Clearly, however, a danger exists, as in the statement "Darwin's law of natural selection implies that natural selection tends to increase fitness...of a population in a given environment by modifying the phenotype (Dodson and Hallam 1977)." The latter statement, as we shall see, follows from the basic axioms of natural selection only under very restrictive hypotheses, and in general will be false. As Gould (1977) has elegantly put it, "If the world displays any harmony and order, it arises only as an incidental result of individuals seeking their own advantage - the economy of Adam Smith transferred to nature." A more precise statement might in fact substitute "genes" for individuals in Gould's description (see Dawkins (1976), Waddington (1957)); but indeed recombination events cause difficulties for even this point of view. One of the goals of this paper is to survey the conditions under which macroscopic statements may be derived from fundamental axioms.

By macroscopic parameters I refer to ones which pertain to the population, and among the most obvious of these are allelic or genotypic frequencies. Both, however, are measured in the currency of the geneticist, whereas *mean fitness* deals with a phenotypic property. Measured in units which are at least in theory directly measurable in the field, mean fitness has hence been a parameter of considerable interest in population genetics.

The mean fitness of the population X^k was formally defined earlier as

$$\bar{w}^k = \sum_i \sum_j p_i^k p_j^k w_{ij}^k. \tag{3}$$

The most important results concerning it have been ones which derive directly from (1) by averaging. For the present, I shall be dealing with a single population and so for simplicity drop the superscript k. Further, and of utmost importance, the fitnesses w_{ij} will be constant.

The most basic result, Fisher's theorem, has many forms and Fisher's own treatment of the subject has been severely criticized (Turner 1970). To begin, I rewrite the equations for gene frequency change (1) as

$$\Delta p_i = p_i(w_{i\cdot} - \bar{w})/\bar{w}, \tag{4}$$

in which $\Delta p_i = p_i' - p_i$. Note that

$$\bar{w} = p_i w_{i\cdot} + (1-p_i)w_{x\cdot}, \tag{5}$$

where $w_{x\cdot}$ is the mean fitness associated with *not* having the allele i at a given locus on a given chromosome. Hence the first equation in (4) may be rewritten

$$\Delta p_i = p_i(1-p_i)(w_{i\cdot} - w_{x\cdot})/\bar{w}. \tag{6}$$

Note further that

$$w_{i\cdot} = p_i w_{ii} + (1-p_i)w_{ix},\ w_{x\cdot} = p_i w_{ix} + (1-p_i)w_{xx}, \tag{7}$$

in which

$$w_{ix} = \sum_{j\neq 1} p_j w_{ij} / \sum_{j\neq i} p_j,\ w_{xx} = \sum_{j\neq i} \sum_{k\neq i} p_j p_k w_{jk} / \sum_{j\neq i} \sum_{k\neq i} p_j p_k.$$

Taking partial derivatives is an extremely precarious operation unless one is careful to specify what is being held constant; I now introduce the notion of the partial derivative $\partial/\partial p_i$ with respect to p_i in the sense that the relative frequencies of the other alleles are unchanged (that is, relative to each other). Note that in the case of two alleles, this becomes a total derivative. Under this convention, clearly

$$\partial w_{ix}/\partial p_i = 0 \quad \text{and} \quad \partial w_{xx}/\partial p_i = 0, \tag{8}$$

so that from (7)

$$\partial w_{i\cdot}/\partial p_i = w_{ii} - w_{ix},\ \partial w_{x\cdot}/\partial p_i = w_{ix} - w_{xx}. \tag{9}$$

Thus

$$\partial\bar{w}/\partial p_i = w_{i\cdot} - w_{x\cdot} + p_i(w_{ii} - w_{ix}) + (1-p_i)(w_{ix} - w_{xx})$$
$$= 2(w_{i\cdot} - w_{x\cdot}); \tag{10}$$

and hence (6) may be rewritten

$$\Delta p_i = p_i(1-p_i)(\partial\bar{w}/\partial p_i)/2\bar{w} \tag{11}$$

which is basically the formulation due to Wright (1949; see also Li (1975). Now note that

$$\Delta(p_i p_j) = \Delta p_i \Delta p_j + p_i \Delta p_j + p_j \Delta p_i .$$

Hence, from (8),

$$\Delta\bar{w} = \sum_i \sum_j w_{ij} \Delta p_i \Delta p_j + 2\sum_i w_{i\cdot} \Delta p_i . \tag{12}$$

Define now the *additive genetic* or *genic variance*

$$V_g = 2\sum_i p_i (w_{i\cdot} - \bar{w})^2 = \sum_i p_i (1-p_i)(\partial\bar{w}/\partial p_i)^2 . \tag{13}$$

(Li 1967a,b; Crow and Kimura 1970; Nagylaki 1977). Then from either (4) or (11),

$$2\sum_i w_{i\cdot} \Delta p_i = V_g/\bar{w}, \tag{14}$$

so that

$$\Delta\bar{w} = V_g/\bar{w} + \sum_i \sum_j w_{ij} \Delta p_i \Delta p_j . \tag{15}$$

If the Δp_i are all small relative to the p_i, then this justifies the approximation

$$\Delta\bar{w} \sim V_g/\bar{w}, \tag{16}$$

which is what is commonly understood as Fisher's theorem: *The rate of change of fitness is approximately proportional to the genic variance of the population.*

The "error" term in (15),

$$\sum_i \sum_j w_{ij} \Delta p_i \Delta p_j$$

is discussed in somewhat more detail elsewhere (Turner 1970). Most important (Mulholland and Smith 1959, Scheuer and Mandel 1959, Kingman (1961a, 1961b) it may be shown that $\Delta\bar{w} > 0$ unless the genic variance is zero.

I note in passing that it follows easily from (4) and (13) that

$$\sum(\Delta p_i)^2/p_i = V_g/2\bar{w}^2, \tag{17}$$

a condition which I shall have occasion to return to later.

That mean fitness increases under a constant selection regime accords nicely with intuition. Moreover, since $\bar{w}$ is bounded above by the largest value of w_{ij}, $\bar{w}$ cannot increase without bound and will eventually reach an equilibrium value at which, from (15), the genic variance must be zero. From (11), Δp_i will be zero for each i, and an equilibrium (not necessarily polymorphic) results. Thus, basically by the method of Liapunov functions, the problem of finding stable equilibria for the process defined by (4) is reduced to finding the local maxima of the mean fitness function $\bar{w} = \sum_i \sum_j p_i p_j w_{ij}$. If there is a unique such maximum, it is globally stable from the interior of the positive orthant; every non-trivial initial set of frequencies will tend to it.

In general, the adaptive *surface* $\bar{w}(p)$ (Wright 1967) may have multiple peaks, but provides nonetheless a convenient paradigm for the evolutionary process. Selection defines in our earlier sense the algorithm by which fitness is maximized. This has the ring of a general truth; however, it is not. The derivation we have given rests firmly on the constancy of the fitness matrix, and so does the result. Moreover, a major frustration of evolutionary theory has been the unsuccessful attempt to find the analogues of this result if the assumption of constancy is relaxed and the "adaptive surface" wobbles. I shall return to this problem in a later section in discussing density-dependent selection, but even more serious problems exist under frequency-dependent selection or when multiple loci are involved (see for example Kimura 1965). For multiple loci and constant fitnesses Ewens (1969 a,b) shows that mean fitness still increases in the absence of epistasis. An excellent and more general discussion of conditions under which the fundamental theorem is still approximately true may be found in Nagylaki (1976), both for the discrete- and continuous-time models.

In the constant fitness case, other macroscopic parameters have also been considered. In the multi-locus situation, Ewens (1976) focuses attention on the gametic variance. For one locus Crow and Kimura (1970), dealing with the continuous-time approximation, argue that natural selection acts "to produce the maximum increase in fitness possible within the restrictions imposed by the amount of genic variance." Unfortunately, their treatment of the constraint is a bit confusing. The side condition which they actually impose is that $\sum(dp_i/dt)^2/p_i = V_g/2$, which is the continuous time analogue of (17) and indeed is a property of the process defined by (4). However, it is not clear why this condition should be imposed *a priori*, which makes the result somewhat circular.

The discrete-time version is even less satisfying, and perhaps therefore is not included in Crow and Kimura's treatment. It would read "Natural selection acts to produce approximately the maximum in fitness subject to the constraint that $\sum(\Delta p_i)^2/p_i = V_g/2\bar{w}^2$." The side constraint is simply the discrete-time analogue of the condition Crow and Kimura use; the disturbing part of the result, however, is the phrase "approximately" which derives from the error term in (12). The precise result is rather that $\sum_i w_{i\cdot}\Delta p_i$ will be maximized subject to the constraints, which is not quite the same as maximizing the fitness increase.

To derive the above result, I use (as do Crow and Kimura) the method of Lagrange multipliers to maximize $\sum_i w_{i.} \Delta p_i$ subject to the side-constraints (2) and $\sum_i (\Delta p_i) = 0$. The standard method is to introduce the Hamiltonian

$$H = \sum_i w_{i.} \Delta p_i + \lambda(\sum_i (\Delta p_i^2)/p_i - V_g/2\bar{w}^2) + \mu(\sum_i \Delta p_i). \tag{18}$$

where λ and μ are multipliers to be determined. Only the Δp_i are allowed to vary. To obtain the maximum, one differentiates H with respect to each Δp_i and equates to 0, yielding

$$0 = \partial H/\partial(\Delta p_i) = w_{i.} + 2\lambda(\Delta p_i)/p_i + \mu. \tag{19}$$

Multiplying through by p_i and summing over i, I obtain

$$0 = \bar{w} + \mu, \tag{20}$$

since $\sum_{i=1}^{n} p_i = 1$. Substituting for μ in (19), I then obtain the equations

$$\Delta p_i = p_i(w_i - \bar{w})/2\lambda.$$

From this, it follows easily that

$$\sum_i (\Delta p_i)^2/p_i = V_g/8\lambda^2;$$

hence from the side condition $2\lambda = \bar{w}$ and so

$$\Delta p_i = p_i(w_i - \bar{w})/\bar{w},$$

which is the same as (4). The result is, I fear, a Pyrrhic victory, and sheds little light on the evolutionary process.

Ginzburg (1977c; see also 1972), in a very novel paper, introduces in the continuous time case the notion of *fitness entropy*,

$$H(p,p^*) = -\sum_{i=1}^{n} p_i^*(\ln(p_i/p_i^*) \tag{21}$$

where $p_i^*(i=1,\ldots,n)$ are the gene frequencies at equilibrium (only alleles are considered which are present at equilibrium). Clearly $H(\underset{\sim}{p},\underset{\sim}{p}^*) > 0$ if $\underset{\sim}{p} \neq \underset{\sim}{p}^*$ and $H(\underset{\sim}{p}^*,\underset{\sim}{p}^*) = 0$. Ginzburg further shows that

$$\frac{dH}{dt} = m-m^*. \tag{22}$$

(m is the malthusian parameter, the continuous analogue of w). Since m is maximized at equilibrium, H is a *Liapunov function:* it decreases during the process of selection, and attains its minimum (zero) at the equilibrium.

Ginzburg introduces a second macroscopic parameter, the *selective delay*, which is directly observable experimentally. As it approaches equilibrium, any population with constant fitnesses will tend towards exponential growth at the exponential rate m^*. Its population size at (large) time t may be shown to be approximately the same as that of a population of the same initial size but already in genetic equilibrium, and which has been growing for τ less time units. The parameter $\tau(p,p^*)$, which is Ginzburg's selective delay, is essentially the time lost as a result of having been initially out of equilibrium. Ginzburg then shows that his two macroscopic parameters are related by

$$H(p,p^*) = w^*\tau(p,p^*), \tag{23}$$

which also enables one to determine the fitness entropy experimentally.

Demetrius (1974), extending his work on age-structured populations, introduces a somewhat different notion of entropy and proves a Fisher-type theorem and other results. This notion has been developed further in Demetrius (1977). Demetrius' main motivation is to deal with the fact that environments are fluctuating unpredictably, so that conclusions based entirely on deterministic models are insufficient. Drawing on recent developments in ergodic theory, he introduces for an equilibrium state the notion of population *entropy*

$$H = -\sum_{j=1}^{n} p_j \ln p_j / \sum_{j=1}^{n} j p_j,$$

where $p_j = \ell_j m_j/\lambda^j$ defines the probability distribution for the age of reproducing individuals in the stationary population. Here, as in standard usage, ℓ_j is the probability of survival to age j, m_j is the number of offspring an individual in age group j contributes at the next time interval, and $\ln\lambda$ is the intrinsic rate of natural increase (the Malthusian parameter). He further introduces the *reproductive potential*

$$\Phi = -\sum_{j=1}^{n} p_j \ln(\lambda^j p_j) / \sum_{j=1}^{n} j p_j,$$

which is related to H by the formula,

$$\ln\lambda = H - \Phi;$$

Demetrius also develops continuous-time analogues. The quantities $E = \ln\lambda$ and ϕ correspond to the notions of free energy and mean energy.

Demetrius (1974) shows that in randomly mating populations in Hardy-Weinberg equilibrium, and under slow selection, the "rate of change of entropy is equal to the genetic variance in entropy minus the genetic covariance of entropy and reproductive potential." He argues (Demetrius 1977) that in fluctuating environments, there will be selection for high entropy (H selection), and that this is an essential alternative to theories based on r- and K-selection alone.

D. Density-dependent selection and coevolution

Attempts to interface mathematically ecology and genetics have been predominantly single species investigations either of density-dependent selection (MacArthur 1962) or of selection in age-structured populations (Norton 1928, Charlesworth and Giesel 1972, Demetrius 1975, Nagylaki 1977, Rocklin and Oster 1976). MacArthur's approach, being based on a continuous time model, is consequently plagued with a number of technical difficulties characteristic of overlapping-generation models, especially the non-isomorphism between allelic and genotypic distribution.

More recent discussions of density-dependent selection (Anderson 1971, Charlesworth 1971, Roughgarden 1971, 1976, 1977, Clarke 1972, and Ginzburg 1977) are based on a discrete time model such as (1), and with fitnesses w^k_{ij} which are functions of the population densities $N^1,\ldots,N^m$. The resultant model takes the form

$$(p^k_i)' = p^k_i w^k_{i\cdot}/\bar{w}^k, \quad (N^k)' = N^k\bar{w}^k \tag{24}$$

where $w^k_{i\cdot} = \sum_j p^k_j w^k_{ij}(N^1,N^2,\ldots,N^m)$ and $\bar{w}^k = \sum p^k_i w^k_{i\cdot}$.

Because the fitnesses are not frequency-dependent it is not difficult to show that the principal transformations of § 3 are valid and (24) may be rewritten

$$\Delta p^k_i = p^k_i(1-p^k_i)(\partial\bar{w}^k/\partial p^k_i)/2\bar{w}, \quad \Delta N^k = N^k(\bar{w}^k - 1) \tag{25}$$

in which partial derivatives are in the same sense as earlier. Recall that in this convention partial differentiation with regard to p^k_i means that the relative frequencies of the other p^k_j, but not their absolute values, are held fixed. Note that, trivially, $\partial\bar{w}^k/\partial p^\ell_i = 0$ for any i if $k \neq \ell$.

From (25), it is clear that any steady-state must be an extreme point for $\bar{w}^k$ on the simplex $p^k_1 +\ldots+ p^k_{n_k} = 1$; it is easily shown (Ginzburg 1977b) that a stable equilibrium must necessarily correspond to a local maximum for each $\bar{w}^k$, subject to changes in the gene frequencies in population k. (N^k is held fixed at its equilibrium value). Indeed (Ginzburg 1977b), since $\bar{w}^k$ is independent of gene frequencies in other populations, at a stable equilibrium each $\bar{w}^k$ is maximized subject to the constraint that $N^1,\ldots,N^m$ are fixed at their equilibrium values; i.e., $\bar{w}^k$ is locally maximized (with value 1) with respect to gene frequency changes in all populations. This, however, is misleading since it is an artifact of the frequency independence and does not carry over to more general situations.

Ginzburg's results are derived in the continuous-time case, but extend easily to the discrete. They correspond to results derived for $m = 1$ (a single population) by Anderson (1971) and Charlesworth (1971), and for arbitrary m but two alleles by Roughgarden (1976, 1977). Local maximization is not a sufficient condition for the stability of an equilibrium. Ginzburg (1976b) introduces the notion of *ecological stability*, essentially the stability of the underlying ecological system with gene frequencies held fixed, and shows that this and fitness maximization provide the necessary and sufficient conditions for stability. Roughgarden (1977), using algorithmic methods, provides a somewhat different but basically equivalent approach in the two-allele case.

The above considerations relate only to the stability of a given equilibrium; virtually nothing has been written concerning the possibility of limit cycle type behavior surrounding unstable equilibria. Other authors have determined sufficient conditions for the existence of an equilibrium in (25), primarily with $m = 1$, but the system is too general to expect the identification of necessary and sufficient conditions. The most ubiquitous conditions has been that the fitnesses w^k_{ij} decrease in N^k; however, there are many situations when one would not wish to impose such a condition, and an equilibrium is still certainly possible in those cases. When fitnesses are assumed to depend on N in a logistic manner, so that

$$w_{ij}(N) = 1 + R_{ij}(K_{ij} - N)/K_{ij}, \tag{26}$$

both Anderson (1971) and Charlesworth (1971) showed in the case $n = 2$, $m = 1$ that a polymorphic equilibrium exists provided that the heterozygote does not possess a K value intermediate between that of the two homozygotes; and that the equilibrium is locally stable if the heterozygote has the highest K value (provided the R values are not too large). They showed further that when the polymorphic equilibrium obtains, the equilibrium value of N is given by

$$\hat{N} = (\hat{p}_1^2 R_{11} + 2\hat{p}_1\hat{p}_2 R_{12} + \hat{p}_2^2 R_{22})/(\hat{p}_1^2 R_{11}/K_{11} + 2\hat{p}_1\hat{p}_2 R_{12}/K_{12} + \hat{p}_2^2 R_{22}/K_{22}) \tag{27}$$

where $\hat{p}_1$, $\hat{p}_2$ are the equilibrium gene frequencies. The formula (27) is easily derived from (26) by taking weighted averages (weighted by genotypic frequencies) and summing. It may be rewritten (Anderson 1971)

$$N = \hat{\bar{R}}/(\widehat{\overline{R/K}}) \tag{28}$$

where the superior bar denotes the weighted average and the hat indicates that values are taken at equilibrium. The formula (28) is obviously also valid for any number of alleles, although the speciation of the stability conditions is a more complicated but straightforward application of Ginzburg's ecological stability criteria plus bounds on the R values.

In the case $m = 1$, $n = 2$, the population will fix for one homozygote or the other, and $\bar{K}$ will be locally maximized, unless $K_{12} > K_{11}$, K_{22}. If $K_{12} < K_{11}$, K_{22} however, which homozygote will win out depends upon the R values and the initial conditions, so that it is not necessarily the case that the phenotype ($\equiv$ genotype) with largest K will win. Moreover, in the overdominant case ($K_{12} > K_{11}$, K_{22}), a balanced polymorphism results and all three genotypes will be sustained in the population; $\bar{K}$ is not maximized at equilibrium. Thus it is not strictly correct to state (Roughgarden 1971) "the selective values...lead in a mild environment to the evolution of phenotypes having a high carrying capacity K at the expense of a low intrinsic rate of increase, r."

What is maximized, if anything? Anderson (1971) shows, in the logistic case ($n=2$, $m-1$), that one may define a function $\hat{N}(p)$ implicitly by $\bar{w}(p, \hat{N}(p)) \equiv 1$, i.e. as the equilibrium population size which would obtain if the gene frequency were maintained at p, and that the equilibrium value $\hat{N} = \hat{N}(p)$ is a local maximum of $\hat{N}$ with respect to $\hat{p}$ (see also Charlesworth 1971). Under these conditions, then, natural selection maximizes equilibrium population size, but this is not a result which carries over to the frequency-dependent case. There are straightforward extensions to more alleles (Ginzburg 1977b) and several interacting species (Ginzburg 1977b, Roughgarden 1976, 1977), but the latter requires additional conditions and is somewhat cumbersome.

More generally, suppose that $\partial\bar{w}^k/\partial N^k < 0$ and the functions $\hat{N}^k(\underset{\sim}{p}^1,\ldots,\underset{\sim}{p}^m)$ can be defined by the equations $\bar{w}^k(\underset{\sim}{p}^1,\ldots,\underset{\sim}{p}^m, N^1,\ldots,N^m) = 1$ for $k=1,\ldots,m$. Then (i) when only the gene frequencies p^k of a single population are allowed to vary, (and densities are held fixed), the critical points of the mean fitness $\bar{w}^k$ and the *conditional equilibrium population size* (Roughgarden 1976) $\hat{N}^k$ for that population coincide; and (ii) $\bar{w}^k$ will be maximized at gene frequency equilibrium as a function of $\underset{\sim}{p}^k$ if and only if $\hat{N}^k$ is. Further, if $\partial\bar{w}^k/\partial N^k > 0$ at equilibrium and the $\hat{N}^k$ are defined as above, point (i) still holds but maximization of $\bar{w}^k$ corresponds to minimization of $\hat{N}_k$ and conversely. These general principles are independent of the number of alleles, and of any assumptions concerning the dependence of fitnesses on densities or frequencies. They are proved by Roughgarden (1976) in the frequency independent case and with $n=2$ in each population, but it is trivial to extend them to the more general case.

The feasibility of defining $\hat{N}(p)$ as above is not restricted to logistic fitnesses; if all fitnesses decrease with N and certain other conditions are imposed on the behavior of the fitnesses near $N=0$ and $N=\infty$, $\hat{N}(p)$ can still be shown to exist. The sufficient conditions can be weakened even further. Moreover, given that $\hat{N}(p)$ is defined, it is straightforward to show that the equilibrium $\hat{N}(p)$ is a local maximum of N with respect to $\hat{p}$. Thus the maximization of population size is a property of the general density-dependent model and not just the logistic one.

Recall that under constant (hence density-independent) selection mean fitness is maximized at gene frequency equilibrium. We have seen that this result extends as well to the case when fitnesses depend on densities ($\bar{w}(p,N)$ is maximized at $p=\hat{p}$ subject to the constraint $N=\hat{N}$), and further that under suitable restraints on the fitnesses (e.g., including that they decrease as N increases) population size will stabilize (so that $\bar{w} = 1$) and population size $\hat{N}$ is the maximum attainable value of N subject to the constraint $\bar{w}(p,N) = 1$. Both of these results are equilibrial in nature, and it is not, for example, the case that $\bar{w}$ increases monotonically to its maximum value of 1. Indeed, the maximum is only with regard to changes in p; $\bar{w}$ is not maximized as a function of both N and p. It is also not the case that N (nor of course $\bar{K}$) increases monotonically, and MacArthur (1962) has pointed out that the problem of finding any quantity which increases is an intricate one. If selection is weak, one may think of the density adjustments as taking place rapidly enough that $\hat{N}(p)$ is essentially attained on the fast time scale; the population settles into a quasi-equilibrium which then tracks changes in p on the slower time scale. In a rough way, then, $\bar{w}$ will increase on the fast time scale to its equilibrium value of 1 and stay approximately constant thereafter while N increases to its maximum. Leon and Charlesworth (1976, 1978) discuss this for one and two species, but it is difficult to make precise. For general selection regimes, it is not correct. Nagylaki (1978) provides a careful discussion of this result for multi-allelic loci in the single species case, both for discrete and continuous time.

Other studies have examined problems not discussed in this section, including the introduction of frequency-dependent fitnesses (Levin 1972, Levin and Udovic 1977, Leon 1974, Roughgarden 1976, 1977), but no general theory exists comparable to what has been described in this section. Indeed, in the frequency dependent case $\bar{w}$ may even be minimized at equilibrium. Further, Roughgarden (1971) performs some interesting simulations of logistic density-dependent evolution in seasonal environments and demonstrates the shift towards "r-selection" which results from such environmental fluctuations. Again, the analytic theory has received very little attention.

In summary, the introduction of density-dependence into fitnesses somewhat weakens the generalizations based on the mechanical approach. Fitness is still seen to be maximized at equilibrium, and indeed under some conditions equilibrium population size is as well. However, the fitness maximization algorithm which is natural selection loses some of its glamour in that neither fitness nor anything else clearly identifiable proceeds monotonically to equilibrium; the algorithm is possibly a round-about one. This means that it is not in general true, as Dodson and Hallam (1977) claim, that "the fitness can be compared with a potential function." Hence approaches which rest upon this statement such as those based on catastrophe theory are invalid.

Worse still, in the frequency dependent case or when interactions between loci are considered, even the equilibrial maximization of fitness need not be achieved. Nonetheless, in the light of Nagylaki's results and those of this section, I choose to regard fitness maximization subject to appropriate constraints as the null hypothesis which, however, fails under strong epistasis or frequency dependence.

E. Metric traits and polygenic inheritance

Despite the fact that inheritance is particulate in nature, it would be foolhardy from a mathematical point of view to include every detail when considering traits controlled at many loci, just as it would not make sense to try to keep track of every molecule in a theory of gases. Continuous approximations are in many cases justified, and need not fall into the category of "blending inheritance."

Several recent theoretical studies of "quantitative inheritance" have been made (Kimura 1965, Slatkin 1970, Cavalli-Sforza and Feldman 1976, Lande 1976, Feldman and Cavalli-Sforza 1977, Rocklin and Oster 1977, and Karlin pers. comm.), resting for the most part on a phenotypic description of the population; however Cavalli-Sforza and Feldman deal with the phenotype-genotype pair, which they coin the *phenogenotype*. I discuss in this section the approach of Slatkin (1970), which serves as a basis for the work of Rocklin and Oster (1977) on the evolution of ecological parameters.

Slatkin's basic model assumes a single phenotypic parameter z, (although a multivariate approach is feasible), distributed in the population according to the frequency distribution $p(z)$, where $\int_{-\infty}^{\infty} p(z)\,dz = 1$. $p(z)$ is assumed to measure the distribution after mating but prior to selection. The key assumption (Slatkin 1970) "is that the distribution of the character in the offspring of a mating depends only on the value (sic) of character in the two parents." This allows the definition of a transformation rule, specified by the conditional probability $L(z;\, z_1, z_2)$ = the frequency distribution of z in the offspring of parents of phenotypes z_1, z_2. The definition of L subsumes the genetic mechanisms of recombination, etc., but not selection. The approach thus has some serious disadvantages, the most critical being that the genotypic variation of a phenotypic class is ignored (Lewontin 1974), so that shifts in the genotypic makeup of a phenotypic class cannot be accounted for. Given that the trait is polygenic, this can be especially damaging. T. Nagylaki (pers. comm.) has pointed out to me that models based on this approach, when compared to the detailed treatment of Kimura (1965) and Lande (1976), can predict an equilibrium genetic variance which is too small by a factor at least as great as the square root of the number of loci. This comment is not applicable directly to Slatkin's basic model, which includes no explicit assumptions concerning the genetical basis of the character. However it does indicate some of the limitations of the approach by showing that for certain specific applications (in particular if recombination is important) the model does not give the correct answer.

With Slatkin's basic framework and assuming random mating, it is straightforward to derive the fundamental model

$$p'(z) = \left(\int_{-\infty}^{\infty}\int_{-\infty}^{\infty} p(z_1)p(z_2)S(z_1)S(z_2)L(z;\, z_1,z_2)\right) / \left(\int_{-\infty}^{\infty} p(z)S(z)\,dz\right)^2 \tag{29}$$

for the distribution $p'(z)$ in the next generation. Here $S(z)$ is the relative probability of survival of individuals of type z, thus purely a phenotypic property.

Rocklin and Oster (1976) generalize Slatkin's model and perform simulations for a variety of mating systems and selection regimes. They extend the model to interactions among species, with special attention to a two-species competition model. Proceeding first from the continuous model but later from the particulate, they introduce the notions of cooperative (Pareto) equilibrium and competitive (Nash) equilibrium and discuss optimization and game-theoretic approaches to evolution. Usually (but not always) this places the discussion of evolution at the level of the species.

F. Strategic analysis and game theory

It is obvious, and some of the drawbacks have been discussed previously, that any approach at the species level is loaded with difficulties. The species does not have a "strategy;" rather it has a pattern which has been molded by natural selection. On the other hand, as we have seen in considering the classical particulate models of inheritance, if fitnesses are either constant or density-dependent but frequency-independent (and if epistasis is not important), the mean fitness of a species will at equilibrium be locally maximized with respect to other evolutionary alternatives within that species. Although this result is not in general true under strong frequency-dependence or epistasis between loci, it is fairly robust in that weak exceptions may be allowed (Kimura 1958, Ewens 1969 a,b, Nagylaki 1976, Ginzburg 1977b). Even if fitnesses depend upon gene frequencies in other species, it may still be shown (Ginzburg 1977b) that the equilibrium is a critical point, although it does not necessarily follow that it is a maximum. The comparable theory has not been developed in the continuous version described in the previous section, but it is reasonable to expect that analogous results will obtain. The development of such a theory would be of great interest.

For those situations where $\bar{w}$ is indeed maximized at equilibrium with respect to gene frequency change within the species, the essentials for either an optimization or game-theoretic approach are present. For coevolutionary problems, since at equilibrium each w^k is maximized with respect to changes in $\underset{\sim}{p}^k$ but not necessarily $\underset{\sim}{p}^\ell\ (k \neq \ell)$, the equilibrium is a competitive (Nash) equilibrium in the sense of game theory with $\bar{w}^k$ representing the payoff to species k. This must be approached with some caution; it holds only for the equilibrium values of the population densities, which are in turn related to the gene frequencies. This means that the theory is incomplete without appropriate side constraints. However under the assumptions of equilibrium for population densities we know not only that $\bar{w}$ is maximal at equilibrium, we actually know its value (one). From this we obtain the needed constraints. Note that in general there will be values of the species densities close to the equilibrium at which $\bar{w}$ will be greater than 1, so that the sense in which $\bar{w}$ is maximized must be precisely understood; that is, the constraints must be always borne in mind.

Further, we have already seen that there are situations in which fitness is not maximized at equilibrium or for which the equilibrium is not even a critical point. It is also the case that any equilibrium-based theory may be unjustifiably restrictive. For this reason, optimization or game theoretic approaches should be invoked only with great caution.

Nonetheless, we have a starting point and the hope is that the exceptional cases can be dealt with by modification of the basic paradigm. Although the game theoretic approach may not be always strictly applicable in the sense discussed in this section (Slobodkin 1964), it nonetheless provides the most consistently valid generalization and a promising way to view evolutionary problems. Even the single species case is a problem of coevolution between species and environment (Lewontin 1961), the latter containing both biotic and abiotic components. Furthermore game theory is far richer in its possibilities than the simple usages given above; as the simplistic restrictions of this paper are relaxed, it is to be hoped that the more sophisticated mathematical theory will help to meet the challenge.

The idea of using game theory to address evolutionary problems is not new (Lewontin 1961; Slobodkin 1964, 1968; Slobodkin and Rapoport 1974; Warburton 1967); nor is a recognition of the inherent problems involved. Slobodkin and Rapoport (1974) view the problem in terms of the species in a game against nature, in which population survival is the goal. They argue (Slobodkin 1968, Slobodkin and Rapoport 1974) that a population will evolve so as to increase its survival potential, and that "this does not involve 'group selection' since the response of an evolutionary unit is the sum of the optimal responses of its component individuals." This seems to me to beg the question, however, since one must still explain why that sum of responses should maximize population survival potential. This objection coincides with the view put forward by Maynard Smith (1976) in his development of the evolutionarily stable strategy (ESS) (see also Maynard Smith 1977, Lawlor and Maynard Smith 1976).

The idea of an ESS is a simple one; what need to be more fully worked out are the formal connections to evolutionary theory, which are clear only in special cases. Maynard Smith (1975) defines an ESS as follows: the strategy $\underline{I}$ is an ESS iff, for all $\underline{J} \neq \underline{I}$,

$$E_I(I) > E_I(J)$$

or (30)

$$E_I(I) = E_I(J) \quad \text{and} \quad E_J(I) > E_J(J),$$

where $E_X(Y)$ is the expected payoff to an individual playing strategy $\underline{Y}$ if the "opponent" plays strategy $\underline{X}$. If a population is made up entirely of individuals playing $\underline{I}$ or $\underline{J}$, and if these are represented in the proportions p and $q = 1-p$, then the average payoff to an individual playing strategy $\underline{I}$ is

$$E_{\cdot}(I) = pE_I(I) + qE_J(I), \tag{31}$$

and that of an individual playing J is

$$E_{\cdot}(J) = pE_I(J) + qE_J(J); \tag{32}$$

the difference between these is,

$$E_{\cdot}(I) - E_{\cdot}(J) = p(E_I(I) - E_I(J)) + q(E_J(I) - E_J(J)) \quad (33)$$

which (if $\underline{I}$ is an ESS) because of (30) will be positive if q is small. Thus if a population is made up almost entirely of individuals playing strategy $\underline{I}$, then any individual playing strategy $\underline{I}$ will have a higher expected payoff than one playing strategy $\underline{J}$. If the payoff function is interpreted as fitness and reproduction is asexual, then a mean higher payoff for $\underline{I}$-playing individuals means that they will increase in the population. The population will then tend to a homogeneous situation where all individuals play strategy $\underline{I}$, and the latter homogeneous situation is *evolutionarily stable* in the sense that the strategy $\underline{J}$ cannot invade.

If there are two strategies $\underline{I}, \underline{J}$, neither of which is evolutionarily stable, then the strategy $\underline{I}$ will be favored provided $E_{\cdot}(I) > E_{\cdot}(J)$, and $\underline{J}$ provided $E_{\cdot}(I) < E_{\cdot}(J)$. Thus from (33) the point

$$\hat{p} = (E_J(I) - E_J(J))/(E_J(I) - E_J(J) + E_I(J) - E_I(I)) \quad (34)$$

represents an equilibrium point. From below $\hat{p}$, the value of p will increase; from above, it will decrease. Except for the possibility of overshoot (which will not occur with the usual definition of fitness), this equilibrium would be stable. Moreover, if one defines the mean fitness

$$\bar{E} = p^2 E_I(I) + pq(E_I(J) + E_J(I)) + q^2 E_J(J), \quad (35)$$

it is easy to show that $\bar{E}$ is locally maximized at $\hat{p}$ in this case. It would, of course, also be maximized locally at $p = 0$ or 1 if either $\underline{J}$ or $\underline{I}$ respectively were an ESS.

The above discussion is based on *pure* strategies: an individual is either an $\underline{I}$-player or a $\underline{J}$-player; and in Maynard Smith's terminology, these strategies "breed true." One could conceive a situation where individuals can play a mixed strategy ($\underline{I}$ with probability x, $\underline{J}$ with probability y), or where a single genotype produces $\underline{I}$ players with probability x and $\underline{J}$ players with probability y. The latter can be reinterpreted in terms of the former, since the genotype can be thought of as playing "the game" through its off-spring. The mixed-strategy situation is considered to some extent by Maynard Smith (1977). The situation is, however, more complicated than discussed there, since there are infinitely many possible equilibria for the species. For simplicity here, I ignore the possibility of the mixed strategy. The equilibrium population may then have a mixture of strategies, but I avoid the confusion caused by referring to this as the population's "mixed strategy."

The ESS has a number of applications to evolutionary problems. In the asexual case just discussed, the population will tend to maximize fitness $\bar{E}$. This is equivalent to the result we obtained in earlier sections in the sexual diploid case, but in the asexual case frequency-dependent fitnesses are permissible. The notion can also be extended to competition between species. In the diploid case, problems arise with the concept of the ESS (Levin and Udovic 1977, Maynard Smith 1977) since strategies do not breed true, but then the general approach can clearly be generalized.

The idea can also be extended in a different way to randomly-mating diploid populations, either outcrossing hermaphroditic or sexual, by consideration of allelic "strategies." An allele basically corresponds to a strategy. If p is the frequency of the $\underline{I}$ allele in the population, then the mean fitness associated with the $\underline{I}$ strategy is $E_{\cdot}(I) = pE_I(I) + (1-p)E_J(I)$, and that of a second strategy (allele) $\underline{J}$ is $E_{\cdot}(J) = pE_I(J) + (1-p)E_J(J)$. In this usage, $E_I(J) = E_J(I)$; the identification of the sexual partner as the "opponent" has certain overtones which I shall not pursue. With this approach, which Maynard Smith does not discuss, the single-locus diploid situation becomes identical to that developed in earlier sections; $\bar{E}$ is maximized at equilibrium and increases monotonically. This is of course the constant fitness case, so that there are no surprises. The approach at the level of the gene (Dawkins 1976) is logically more consistent than that at the level of the individual; it avoids the paradoxes associated with diploid inheritance by regarding the diploid individual as simply one of the arenas of interaction of the genes. However, the major obstacle to its acceptance is that it does now view evolution at the level of organization at which we as organisms would prefer to do so.

Note that the population need not attain an ESS for a variety of reasons. Indeed, there need not even exist an ESS, or one may exist but not breed true (if it is heterozygous). In the case considered earlier in which two strategies $\underline{I}$ and $\underline{J}$ exist and neither is an ESS, the population will tend to a mixture of strategies which may be thought of as corresponding to a mixed strategy; but that mixed strategy will not in general meet Maynard Smith's criteria for an ESS (Maynard Smith 1977). The mixture of strategies maximizes fitness, as does an ESS; and thus I believe that fitness maximization is a more general paradigm than the ESS. Nonetheless, this alternative way of looking at the problem may be useful.

G. Applications

The viewpoint that emerges from the preceding is that under fairly general conditions, a population will evolve to maximize mean fitness, subject to the constraint that population densities and the genetics of interacting species are held constant. An analogous point of view has formed the basis of studies of optimal life history strategies (Stearns 1976), although it has been often misused. Levin, Cohen, and Hastings (unpublished) have used a similar approach to study optimal dispersal and dormancy strategies, and Rocklin and Oster (unpublished) have examined plant-pollinator interactions. It should be noted that these uses are steady-state in nature and assume that organisms have adapted to their present environments as well as possible. It is an important restriction that when the environment varies too rapidly for this to occur, the paradigm of fitness maximization collapses.

Species competition models have received considerable attention, especially as regards character displacement (MacArthur and Levins (1964, 1967); Fenchel and Christiansen (1977); Slatkin and Lande

(1976); Roughgarden (1976); Lawlor and Maynard Smith (1976); Maynard Smith (1977). The approaches of Roughgarden (1976) and of Lawlor and Maynard Smith (1976) to this problem are similar, but have some fundamental differences. Both consider Lotka-Volterra type growth for two species, Roughgarden in discrete time and Lawlor and Maynard Smith in continuous time. In both, certain parameters are variable, subject to constraints. The models differ in other details, some non-essential (Roughgarden considers a continuous resource spectrum, Lawlor and Maynard Smith a discrete one). Both implicitly utilize the notion of fitness optimization to derive optimal resource utilization patterns. The results have the same general flavor, but differ in specifics due to differences in the models.

Lawlor and Maynard Smith (1975) write an equation for each species of the form

$$dN^k/dt = N^k \phi^k(\underset{\sim}{a}^k; N^1,\ldots,N^m), \tag{36}$$

in which $\underset{\sim}{a}^k$ is the vector of evolutionary parameters, which is determined by the allelic frequencies. The set Ω of possible $\underset{\sim}{a}$ vectors defines the fitness set of Levins (1969); Rocklin and Oster (1976) show further that all that is of interest in this set is the portion of the boundary known in game theory as the Pareto set. This need not concern us here. Since ϕ^k represents the (Malthusian) mean fitness for species k, the general approach is to maximize ϕ^k as a function of $\underset{\sim}{a}$, subject to the constraint that the values $N^1,\ldots,N^m$ are fixed at their equilibrium values. The first-order maximization conditions together with the equations $\phi^k = 0$ provide a set of independent equations equal in number to the number of independent variables, so the problem seems properly posed and the general approach an extremely promising one. Indeed, even when it is only known that ϕ^k has an *extremum* for the equilibrium, the technique can be applied in part since the first-order equations are identical.

I do not discuss the results of this analysis here, for which the reader is referred to Lawlor and Maynard Smith (1975) or the similar results of Roughgarden (1976). It must be emphasized, as those authors realize, that the utilization of this approach makes certain implicit assumptions about the underlying genetics. Roughgarden has pointed out that especially in the consideration of interspecific interactions, the most crucial assumption (frequency independence) is likely to be violated. Despite its promise, there are thus still major difficulties with the application of the approach. Slatkin (1976) provides some discussion of such applications, and interprets them as implicitly based upon selection on modifier loci (see also Karlin and McGregor 1974). He provides an alternative paradigm for the frequency-dependent case (equilibration of fitnesses of the various phenotypes), and details the conditions under which it is valid. In recent work (Slatkin pers. comm.), he has successfully applied this approach to problems of character displacement and derived new conclusions.

H. Summary

Optimization theory and game theory have been frequently applied to questions of adaptation (Levins 1968, Cody 1974), often too glibly. On the other hand, an approach to evolution which is too involved with the detailed mechanics of the genetic process is both mathematically impractical and virtually useless in that it is embarrassingly rich in parameters. What is needed is an evolutionary theory which blends the best of both approaches, couched in terms of macroscopic parameters of the population but firmly based on the known mechanisms of Darwinian natural selection. It may well be, as Jacob (1977) has discussed, that the best that we can hope for such a theory is that it be explanatory rather than predictive.

If fitnesses are constant and controlled at a single locus, natural selection will act to maximize mean fitness. This result extends to density-dependent fitnesses in that at equilibrium, fitness is maximized subject to the constraint that population density is at its equilibrium value. When extended to interacting populations the result continues to apply: each species maximizes its equilibrial mean fitness subject to the constraints that all other species are genetically fixed and population sizes in all species are fixed. Although strong frequency dependence or epistasis can vitiate the main result, it still appears fairly robust as a general conclusion.

Moreover, given that fitness is maximized subject to the stated constraints, a coevolutionary equilibrium corresponds to the game-theoretic notion of a Nash equilibrium. Game-theoretic approaches (Rocklin and Oster 1976, Maynard Smith 1975) or equivalent ones (Roughgarden 1976) have therefore been advanced recently by several authors to address the evolution of ecological parameters, and the various attempts have as their core the basic principle stated above.

Fitness optimization does not apply in all cases, and it is to be hoped that a more refined theory will eventually be able to deal more satisfactorily, for example, with frequency-dependent fitnesses, interactions between loci, and changing environments. Until such time, considerable caution must be exercised. Lewontin (1977) points out that "Adaptation, for Darwin, was a process of becoming rather than a state of final optimality." Environments change, not only due to extrinsic causes but also as a result of adaptation already effected. The evolutionary response is to the present, and carries with it no guarantee that any sort of optimum will be attained. Evolution is something which simply happens, rather than a calculated, far-seeing program for optimization.

The failure to make this distinction has led to errors both by the practitioners of evolutionary theory, and by those who have been critica of them. Many, such as Peters (1976), have criticized aspects of the theory as being tautologous; and it is especially tempting to level such criticism at the popular bastardization of evolutionary thought, "Since natural selection is defined as survival of the fittest, it is clear that those that have survived are indeed fittest." This statement, however, is not a tautology, as can be seen from the fact that it is not always correct. Yet it is true, as Lewontin (1977) argues, that "the adaptationist program is so much a part of the vulgarization of Darwinism that an increasing amount of evolutionary theory consists in the uncritical application of the program to both manifest and postulated traits of organisms." Many of the mysteries of nature become clearer when one accepts that seemingly maladaptive bahavior may be just that.

To raise these caveats is not to negate the usefulness or power of optimality arguments. But the distinction Lewontin has emphasized between becoming and being optimal must not be ignored. If optimality or game theoretic approaches are to be invoked, then they must be justified as the inescapable consequences of the becoming, and this will be possible only if additional assumptions have been made (e.g. concerning frequency dependence, epistasis, temporal variation, and the attainment of equilibrium). By the process of justification, the optimizer is forced to lay bare hidden assumptions, and ambiguities concerning what to optimize are stripped away. This review article has been intended as a catalog of situations where such justification is possible.

Acknowledgements: It is a pleasure to acknowledge the support of NSF Grants GP33031 and MCS77-01076 and the very helpful comments of Lloyd Demetrius, Alan Hastings, Tom Nagylaki, George Oster, and Monty Slatkin. The opinions expressed in this paper often differed markedly from ones those colleagues would have deemed prudent, and I claim no endorsement from them.

References

Anderson, W.W.: Genetic equilibrium and population growth under density-regulated selection. Amer. Natur. 105:489-498. (1971).

Asmussen, M.A., Feldman, M.W.: Density-dependent selection I: A stable feasible equilibrium may not be attainable. J. Theor. Biol. 64:603-618. (1977).

Cavalli-Sforza, L.L., Feldman, M.W.: Evolution of continuous variation: Direct approach through joint distribution of genotypes and phenotypes. Proc. Nat. Acad. Sci. 73:1689-1692. (1976).

Charlesworth, B.: Selection in density-regulated populations. Ecology 52:469-474. (1971).

Charlesworth, B., Giesel, J.T.: Selection in populations with overlapping generations IV. Fluctuations in gene frequency with density-dependent selection. Amer. Natur. 106:402-411. (1972).

Clarke, B.C.: Density-dependent selection. Amer. Natur. 106:1-13. (1972).

Cody, M.: Optimization in ecology. Science 183:1156-1164. (1974)

Crow, J.E., Kimura, M.: An Introduction to Population Genetics Theory. New York: Harper and Row, 1970, 591 + xiv pp.

Dawkins, : The Selfish Gene. New York: Oxford University Press, 1976, 224 + xi pp.

Demetrius, L.: Demographic parameters and natural selection. Proc. Nat. Acad. Sci., USA 71:4645-4647. (1974).

Demetrius, L.: Natural selection and age-structured populations. Genetics 79:539-544. (1975).

Demetrius, L.: Measures of fitness and demographic stability. Proc. Nat. Acad. Sci., USA 74:384-386. (1977).

Dodson, M.M.,Hallam, A.: Allopatric speciations and the fold catastrophe. Amer. Natur. 111:415-433. (1977).

Ewens, W.J.: Remarks on the evolutionary effect of natural selection. Genetics 83:601-607. (1976).

Ewens, W.J.: A generalized fundamental theorem of natural selection. Genetics 63:531-537. (1969a).

Ewens, W.J.: Mean fitness increases when fitnesses are additive.

Nature 221:1076. (1969b).
Feldman, M.W., Cavalli-Sforza, L.L.: The evolution of continuous variation II. Complex transmission and assortative mating. Theor. Pop. Biol. 11:161-181. (1977).
Fenchel, T.M., Christiansen, F.B.: Selection and interspecific competition. In Symposium on the measurement of selection in natural populations. (ed. F.B. Christiansen and T. Fenchel). Lecture Notes in Biomathematics. Heidelberg: Springer-Verlag. (In press).
Fisher, R.A.: The Genetical Theory of Natural Selection. Oxford: Clarendon Press, 1930.
Ford, E.B.: Ecological Genetics. London: Methuen; New York: J. Wiley, 1964.
Ginzburg, L.R.: Diversity of fitness and generalized fitness. J. Gen. Biol. 33:77-81. (1972).
Ginzburg, L.R.: The equilibrium and stability for n alleles under the density-dependent selection. J. Theor. Biol. (In press). (1977a).
Ginzburg, L.R.: Local considerations of polymorphisms for populations coexisting in stable polymorphisms. J. Math. Biol. (In press). (1977b).
Ginzburg, L.R.: A macroequation of natural selection. J. Theor. Biol. (In press). (1977c).
Gould, S.J.: Ever Since Darwin. New York: Norton, 1977.
Jacob, F.: Evolution and tinkering. Science 196:1161-1166. (1977).
Jayakar, S.D.: A mathematical model for interaction between gene frequencies in a parasite and its host. Theor. Pop. Biol. 3: 210-238. (1970).
Karlin, S., McGregor, J.: Towards a theory of the evolution of modifier genes. Theor. Pop. Biol. 5:59-103. (1974).
Kimura, M.: On the change of population fitness by natural selection. Heredity 12:145-167. (1958).
Kimura, M.: Some recent advances in the theory of population genetics. Jap. J. Human. Genet. 10:43-48. (1965).
Kingman, J.F.C.: A matrix inequality. Quart. J. Math. 12:78-80. (1961a).
Kingman, J.F.C.: A mathematical problem in population genetics. Proc. Cambridge Phil. Soc. 57:574-582. (1961b).
Krebs, C.J., Keller, B.C., Tamarin, R.H.: *Microtus* population biology: Demographic changes in fluctuating populations of *M. ochrogaster* and *M. pennsylvanicus* in southern Indiana. Ecology 50:587-607. (1969).
Lande, R.: The maintenance of genetic variability by mutation in a polygenic character with linked loci. Genet. Res. 26:221-235. (1976).
Lawlor, J.R., Maynard Smith, J.: The coevolution and stability of competing species. Amer. Natur. 110:79-99. (1976).
Leon, J.A.: Selection in contexts of interspecific competition. Amer. Natur. 108:739-757. (1974).
Leon, J.A., Charlesworth, B.: Ecological versions of Fisher's fundamental theorem of natural selection. Advances in Applied Probability 8:639-641. Symposium on Mathematical Genetics. Liverpool, 5-6 April 1976.
Leon, J.A., Charlesworth, B.: Ecological versions of Fisher's fundamental theorem of natural selection. M.S. (1977).
Levin, S.A.: A mathematical analysis of the genetic feedback mechanism. Amer. Natur. 106:145-164. (1972). (Erratum 1973. Amer. Natur. 107:302).
Levin, S.A., Udovic, J.D.: A mathematical model of coevolving populations. Amer. Natur. 111:657-675. (1977).
Levins, R.: Evolution in changing environments. Princeton:

Princeton University Press, 1968, 120 + ix pp.
Lewontin, R.C.: Evolution and the theory of games. J. Theor. Biol. 1:382-403. (1961).
Lewontin, R.C.: The Genetic Basis of Evolutionary Change. New York: Columbia University Press, 1974, 346 + xiii pp.
Lewontin, R.C.: Adaptation. In Enciclopedia Einaudi Turin 1:198-214. (1977).
Lewontin, R.C.: Fitness, survival, and optimality. In Analysis of Ecological Systems. (ed. D.J. Horn, R. Mitchell and G.R. Stairs). Columbus, Ohio: Ohio University Press, 1978.
Li, C.C.: The stability of an equilibrium and the average fitness of a population. Amer. Natur. 89:281-295. (1955).
Li, C.C.: Fundamental theorem of natural selection. Nature 214: 505-506. (1967a).
Li, C.C.: Genetic equilibrium under selection. Biometrics 23: 397-484. (1967b).
MacArthur, R.H.: Some generalized theorems of natural selection. Proc. Natl. Acad. Sci. 48:1893-1897. (1962).
MacArthur, R.H., Levins, R.: Competition, habitat selection, and character displacement in a patchy environment. Proc. Natl. Acad. Sci. 51:1207-1210. (1964).
MacArthur, R.H., Levins, R.: The limiting similarity, convergence, and divergence of coexisting species. Amer. Natur. 101:377-385. (1967).
Marr, D.: The processing of visual information. In Lectures on Mathematics in the Life Sciences X. Some Mathematical Problems in Biology 9. (ed. S.A. Levin). Providence, R.I.: American Mathematical Society, 1978. (In press).
Maynard Smith, J.: Evolution and the theory of games. Amer. Sci. 64:41-45. (1976).
Maynard Smith, J.: Evolution and the theory of games. In Mathematics in the Life Sciences. Lecture Notes in Biomathematics. (ed. W. Matthews). Berlin, New York: Springer-Verlag. (1977).
Mode, C.J.: A mathematical model for the coevolution of obligate parasites and their hosts. Evolution 12:158-165. (1958).
Mulholland, H.P., Smith, C.A.B.: An inequality arising in genetical theory. Amer. Math. Monthly 66:673-683. (1959).
Nagylaki, T.: The evolution of one- and two-locus systems. Genetics 83:583-600. (1976).
Nagylaki, T.: Selection in one- and two-locus systems. Lecture Notes in Biomathematics. Heidelberg: Springer-Verlag, 1977.
Nagylaki, T.: The dynamics of density- and frequency-dependent selection. (M.S., 1978).
Norton, H.T.J.: Natural selection and Mendelian variation. Proc. Lond. Math. Soc. 28:1-45. (1928).
Oster, G.: Stochastic behavior of deterministic models. pp. 24-37 In Ecosystem Analysis and Prediction. (ed. S.A. Levin). Philadelphia: SIAM.
Peters, R.H.: Tautology in evolution and ecology. Amer. Natur. 110:1-12. (1976).
Pimentel, D.: Animal population regulations by the genetic feedback mechanism. Amer. Natur. 95:65-79. (1961).
Pimentel, D.: Population regulation and genetic feedback. Science 159:1432=1437. (1968).
Pimentel, D., Levin, S.A., Soans, A.B.: On the evolution of energy balance in some exploiter-victim systems. Ecology 56:381-383. (1975).
Rocklin, S., Oster, G.: Competition between phenotypes. J. Math. Biol. 3:225-262. (1976).
Roughgarden, J.: Density-dependent natural selection. Ecology 52: 453-468. (1971).

Roughgarden, J.: Resource partitioning among competing species. A coevolutionary approach. Theor. Pop. Biol. 9:388-424. (1976).
Roughgarden, J.: Coevolution in ecological systems II. Results from "loop analysis" for purely density-dependent coevolution. In Symposium on the Measurement of Selection in Natural Populations. Lecture Notes in Biomathematics. (ed. F.B. Christiansen and T. Fenchel). Heidelberg: Springer-Verlag, 1977. (In press).
Scheuer, P.A.G., Mandel, S.P.H.: An inequality in population genetics. Heredity 13:519-524. (1959).
Schutz, W.M., Usanis, S.A.: Intergenotypic competition in plant populations II. Maintenance of allelic polymorphisms with frequency-dependent selection and mixed selfing and random mating. Genetics 61:875-891. (1969).
Slatkin, M.: Selection and polygenic characters. Proc. Nat. Acad. Sci. 66:87-93. (1970).
Slatkin, M.: On the equilibration of fitnesses by natural selection (MS). (1976).
Slatkin, M., Lande, R.: Niche width in a fluctuating environment-density independent model. Amer. Natur. 110:31-55. (1976).
Slobodkin, L.B.: The strategy of evolution. Amer. Sci. 52:342-357. (1964).
Slobodkin, L.B.: Toward a predictive theory of evolution. pp. 187-205 In Population Biology and Evolution. (ed. R.C.Lewontin). Syracuse: University of Syracuse Press, 1968.
Slobodkin, L.B., Rapoport, A.: An optimal strategy of evolution. Quart. Rev. Biology 49:181-200. (1974).
Smale, S.: On the differential equations of population in competition. J. Math. Bio. 3:5-7. (1976).
Stearns, S.C.: Life history tactics: A review of the ideas. Quart Rev. Biol. 51:1207-1210. (1976).
Stewart, F.M.: Evolution of dimorphism in a predator-prey model. Theor. Pop. Biol. 2:493-506. (1971).
Turner, J.R.G.: Changes in mean fitness under natural selection. pp. 32-77 In Mathematical Topics in Population Genetics. (ed. K. Kojima). Berlin: Springer-Verlag, 1970.
Waddington, C.H.: The Strategy of the Genes. London: Allen and Unwin, 1957.
Warburton, F.E.: A model of natural selection based on a theory of guessing games. J. Theor. Biol. 16:78-96. (1967).
Wright, S.: Adaptation and selection. pp. 365-389 In Genetics, Paleontology, and Evolution. (ed. G.L. Jepson, G.G. Simpson, and E. Mayr). Princeton, New Jersey: Princeton University Press, 1949.
Wright, S.: Classification of the factors of evolution. Cold Spring Harbor Symp. Quant. Biol. 20:16-24. (1955).
Wright, S.: "Surfaces" of selective value. Proc. Nat. Acad. Sci. 58:165-172. (1967).
Wright, S.: Evolution and the Genetics of Populations II. Chicago: The University of Chicago Press, 1969.
Yu, P.: Some host parasite genetic interaction models. Theor. Pop. Biol. 3:347-357. (1972).

Coevolution in Ecological Systems III. Coadaptation and Equilibrium Population Size

J. Roughgarden

A. Introduction

The effort toward uniting evolutionary theory with ecological theory has been plagued from the beginning by a conflict between the optimization criterion in evolutionary theory and criteria from ecological theory. In evolutionary theory, natural selection is viewed as producing traits which maximize the fitness of the individuals carrying them relative to the fitness of individuals with alternative traits. The criterion of fitness maximization is often used uncritically and incorrectly in evolutionary theory, especially in sociobiological contexts, but it can sometimes be justified. One of many examples where it is justifiable is discussed below. The problem to be discussed in this paper, however, does not lie with the criterion of fitness maximization in itself, but with the fact that optimum strategies computed according to the fitness criterion do not generally coincide with optimum strategies determined from various ecological optimization criteria. Specifically, one common type of optimum ecological strategy maximizes the abundance of a population. In ecology it is natural to ask what strategies of resource use and allocation will lead to the highest population size, i.e., what predation rate leads to optimum harvesting of the prey, how much energy should prey allocate to predator defense, etc. For these questions, the criterion is the maximization of population size. The optimum predation rate is that which yields the highest sustainable yield, and analogously for the optimum prey defense. However, a conflict arises whenever the strategies which maximize fitness fail to coincide with the strategies which maximize the ecological criterion of interest.

This conflict cannot be resolved by compromise. Whenever the strategies from evolutionary criteria do not coincide with those from ecological criteria, then the ecological criteria fail. The appropriateness of ecological optimization criteria is derivative upon evolutionary criteria. What we can do, however, is to clarify the relationship of strategies which maximize fitness with strategies which maximize population size. Specifically, we can first determine the value of some trait, say X_s, which maximizes the fitness maximization as $\hat{X}_s$. Next we can calculate the quantity $(d\hat{N}_s/dX_s)|_{\hat{X}_s}$, where $\hat{N}_s$ is the equilibrium population size of species-s. This derivative describes how the equilibrium population size of species-s would vary as the trait is varied about its evolutionary equilibrium position. For example, suppose $\hat{X}_s$ refers to the amount of energy allocated by a prey animal toward predator defense at evolutionary equilibrium. Then if $d\hat{N}_s/dX_s$ is positive, it means that prey defense evolves to a level which is below that which would maximize the prey abundance. Similarly, if $d\hat{N}_s/dX_s$ is negative, it means that prey defense evolves to a level which is higher than that which would maximize prey abundance. Thus determining the above quantity helps to clarify the differences between strategies which maximize the evolutionary criterion of fitness and those which maximize the ecological criterion of population abundance.

This paper develops a general and largely graphical theory to enable calculation of the quantities $d\hat{N}/dX$ in coevolutionary models. The theory is illustrated in detail with two examples, one of predator-prey coevolution and one for the character displacement of two competitors. It is an extension to interspecific frequency dependence of my earlier paper on purely density-dependent coevolution (Roughgarden 1977).

B. Fitness Maximization in Coevolutionary Theory

The traditional topics in coevolution lead to selection pressures which involve interspecific frequency dependence. This feature of coevolution is extremely important and must be treated with care.

A typical coevolutionary population genetic model is one of the form

$$\Delta N_s = (\bar{W}_s - 1)N_s \qquad s = 1 \,..\, S \tag{1a}$$

$$\Delta p_s = (p_s(1 - p_s)/2\bar{W}_s)\,\partial\bar{W}_s/\partial p_s \tag{1b}$$

where $\bar{W}_s$ is the mean fitness in species-s. There is variation at one locus with two alleles in each species, and there is no intra-specific frequency dependence. More specifically, the fitness of genotype A_iA_j in species-s is of the form

$$W_{s,ij} = W_{s,ij}(N_1 .. N_S,\ p_s .. p_{k\neq s} .. p_S) \tag{2}$$

that is, the fitness of a genotype in species-s depends on *all* the population sizes and on the gene frequencies in all other populations except p_s itself. These fitness functions thus include density dependence and inter-specific frequency dependence, but not intra-specific dependence.

As a concrete example of a typical coevolutionary population genetic model, consider the coevolution of a predator and its prey. We form the coevolutionary model after the pattern of the following familiar predator-prey model in ecology

$$\begin{aligned} \frac{dN_1}{dt} &= \frac{r(K-N_1)N_1}{K} - aN_1N_2 \\ \frac{dN_2}{dt} &= abN_1N_2 - dN_2 \end{aligned} \tag{3}$$

Suppose that the prey has genetic variation for predator defenses and that the predators have genetic variation for overcoming prey defenses. Let $a_{ij,kl}$ be the predation rate of genotype A_kA_l in *the predators* against genotype A_iA_j in the prey. Then we are led to fitness functions of the form

$$W_{1,ij} = 1 + r_{ij} - \frac{r_{ij}}{K_{ij}} N_1 - \left[a_{ij11} p_2^2 + a_{ij,12} 2p_2(1-p_2) + a_{ij,22}(1-p_2)^2 \right] N_2 \tag{4a}$$

$$W_{2,kl} = \left[p_1^2 a_{11,kl} + 2p_1(1-p_1) a_{12,kl} + (1-p_1)^2 a_{22,kl} \right] b_{kl} N_1 - d_{kl} + 1 \tag{4b}$$

Note that the fitnesses are density dependent and exhibit interspecific frequency dependence. The question we now face is whether we are entitled to use strategy reasoning to predict the outcome of selection in such a complicated situation. To answer this question, we present a population genetic result in the following paragraphs.

A coevolutionary equilibrium (stable or unstable) satisfies, for every i, one of the following:

$$\bar{W}_i = 1, \quad \partial \bar{W}_i / \partial p_i = 0 \tag{5a}$$

$$W_{i,11} = 1, \qquad \hat{p}_i = 1 \tag{5b}$$

$$W_{i,22} = 1, \qquad \hat{p}_i = 0 \tag{5c}$$

In (5a) there is polymorphism, and in (5b) and (5c) there is fixation of an allele. Equilibria of type (5b) and (5c) are called *boundary equilibria*.

We also associate with the coevolutionary model another model which is called the "pure ecological model". This model is obtained formally from (1) by taking the p's as constants, leaving us with the equations for the variable population sizes.

$$\Delta N_i = (\bar{W}_i - 1) N_i \tag{6}$$

The set of equilibrium communities which results from this model for all fixed gene frequencies comprises the set of communities which are evolutionarily possible.

With this terminology, we can state the main population genetic result for coevolution with interspecific frequency dependence.

Result:

Assume that the fitnesses have density-dependence and interspecific frequency-dependence, and that the associated pure ecological model has a unique locally stable equilibrium for any set of fixed p's. Then,

(a) An equilibrium point of the coevolutionary model, $(\hat{p}_1 .. \hat{p}_S, \hat{N}_1 .. \hat{N}_S)$ where every species is fixed (i.e., $\hat{p}_i = 0,1$ for all i) is locally stable if, and only if, the mean fitness in each species is maximized with respect to the gene frequency in that species.

(b) An equilibrium point of the evolutionary model, $(\hat{p}_1..\hat{p}_S, \hat{N}_1..\hat{N}_S)$, in which one or more of the species is polymorphic, is not necessarily stable if the mean fitness is maximized in every species, nor is it necessarily unstable if the mean fitness is minimized in some species. See, for example, Levin and Udovic (1977) for the demonstration of this result with two populations.

This result is much weaker than its counterpart for purely density-dependent coevolution (Roughgarden 1977). With pure density-dependent coevolution, fitness is also maximized at polymorphic equilibria. If an equilibrium involves a polymorphism in one or more species, there is simply no general intuitive criterion to determine its stability. The polymorphic equilibria can always be located using formula (5a). But the stability must be settled in each example by detailed reference to the eigenvalues of the stability matrix at each equilibrium point. At present there is no general optimization principle which determines stability when polymorphism is involved.

It is of fundamental interest to know whether fitness is maximized at equilibrium. If it is, then we are entitled to use strategy reasoning to determine what kinds of traits will result from evolution. Part (a) of the above result tells us that we are entitled to use strategy reasoning at least some of the time. It says that there exist situations where coevolution with interspecific frequency dependence does lead to fitness maximization in each species. Furthermore, part (a) asserts that a community configuration in which fitness is maximized in each species is certain to be coevolutionarily stable provided that this configuration is realized without invoking polymorphism in the species involved.

Part (a) of the above result provides a connection with the concept of an "evolutionarily stable strategy" or ESS introduced by Maynard Smith (1974). An ESS is defined as a strategy such that if the population consists entirely of individuals with the ESS, then no mutant allele producing some slightly different strategy can increase. The criterion for an ESS is thus a criterion for the stability of a boundary equilibrium. Part (a) above shows that a community configuration in which fitness is maximized within each of the species is the community analogue of an ESS. A much stronger concept exists for purely density-dependent coevolution, but with interspecific frequency-dependence we must be satisfied with a justification for strategy reasoning in terms of the stability of boundary equilibria.

Part (b) in the above result is also interesting. It indicates a generally destabilizing role for polymorphism in traits which govern species interactions. It is possible for polymorphism to destabilize a strategy where fitness is maximized in every species and which would certainly be stable if attained without recourse to polymorphism. It is also possible for polymorphism to stabilize strategies which minimize fitness.

C. The Routes by Which Evolution in a Species Influences its Own Abundance

Evolution within a species influences its own abundance in two qualitatively different ways. The first way is by a *direct ecological route.* Evolution within a species causes changes in the *abundance* of other species. These changed abundances in the other species feed back upon the abundance of the species whose evolution is under consideration. The distinguishing feature of this direct ecological route is that it is mediated by changes in the abundance of the other species and does *not* involve any evolutionary changes in the other species. If the other species in the community are evolutionarily static or evolutionarily slow, then the effect of evolution within a species on its own abundance is given solely by the effects mediated through this direct ecological route.

The other route involves evolutionary changes in the other species which are induced by the evolution in the species under consideration. A species may evolutionarily influence its own abundance by causing evolutionary change in the *phenotypes* of other species. These changed phenotypes in the other species in turn influence the abundance of the species whose evolution is under study. The distinguishing feature of this evolutionary route is that it is mediated by phenotypic changes in the other species. If the evolution of the species can occur in the same time scale as the ecological changes in the system, then both routes must be considered together.

Formally, we may express these distinctions with the following formula:

$$\frac{d\hat{N}_s}{dX_s} = \frac{\partial \hat{N}_s}{\partial X_s} + \sum_{i \neq s} \frac{\partial \hat{N}_s}{\partial X_i} \frac{\partial \hat{X}_i}{\partial X_s} \qquad (7)$$

The quantity $d\hat{N}_s/dX_s$ may be termed the *total effect* on the abundance of species caused by varying the trait in species-s slightly away from its equilibrium value. The term $\partial \hat{N}_s/\partial X_s$ is called the *direct effect;* this represents the direct ecological route. The remaining terms, all of which are included in the summation over $i \neq s$, represent *evolutionary reciprocation,* i.e., the changes in $\hat{N}_s$ mediated by evolution induced in the remaining species by the variation of X_s.

The formula above is misleadingly simple, because each quantity in it, e.g., $\partial \hat{N}_s/\partial X_s$, is determined by taking account of all the connections between the species. Nonetheless, the formula is helpful in organizing the issues and in classifying the qualitatively different causal processes which enter into the overall problem of coevolution.

A final note is that the presence of interspecific frequency dependence greatly influences the properties of both the direct effect and of the evolutionary reciprocation. It can be shown that in purely density-dependent evolution, the terms representing evolutionary reciprocation are always of lower order than the term representing the direct effect. In purely density-dependent coevolution the total effect is dominated by the direct ecological effect, and the terms for the evolutionary-reciprocation can be ignored for small variations of X_s around $\hat{X}_s$.

D. The Theory of Coadaptation

Natural selection in interacting species causes the evolution of traits in each species which are adaptations to the presence of other species. This section explains the ecological consequences of coadaptation, and it explains how the presence of coadaptation influences the ecology of the species which are coadapted to one another. Here we abandon explicit reference to gene frequencies and take a strategy reasoning approach. This is the actual context in which the theory is useful.

We will assume there is a *trait* in each species whose value is X_s for species-s. For example, X_s could represent the amount of energy that a prey allocates into defensive armor, or X_s could be the amount of time per 24 hours that a predator actually searches for prey, etc. The interaction coefficient between the species are functions of these traits. For example, the standard predator-prey coefficient, a, in the predator-prey model is a function of both X_1 and X_2; i.e., $a = a(X_1, X_2)$. Thus the interaction coefficients between the species, and hence ultimately their equilibrium population sizes, are influenced by the coevolution of these traits. The interaction coefficients are *jointly* controlled in an evolutionary sense by the two interacting species.

We will determine the optimum value of X_s in each species by maximizing the fitness in each species. We are justified in doing this because, as discussed in a preceding section, the community configuration associated with fitness maximization is coevolutionarily stable provided the configuration occurs at a genetic boundary equilibrium. That is, if every species is fixed for an allele which produces a value of the traits X_s, which maximizes the fitness in species-s, then a mutant allele which produces a slightly different value of X_s cannot increase. Our problem is to discover the ecological implications of satisfying the criterion that the value of the trait, X_s, in each species maximizes the fitness, W_s, in that species.

A strategy model is not itself a dynamic model, although it can be used to predict the equilibrium configuration of a dynamic model. The strategy model is defined to be the set of S expressions for the fitness in each species.

$$W_s(X_1..X_S,\ N_1..N_S) \qquad s = 1 .. S \tag{8}$$

The set of traits representing the coevolutionarily stable equilibrium, $(\hat{X}_1..\hat{X}_S)$, is defined as the set of traits such that W_s is maximized with respect to X_s at $X_s = \hat{X}_s$ subject to the constraint that $W_s = 1$, for all s. Furthermore, the equilibrium population sizes, $\hat{N}_i(X_1..X_S)$, are defined implicitly by the equation

$$W_s = 1 \qquad s = 1 .. S \tag{9}$$

These equilibrium population sizes must be positive and unique for all values of the traits under consideration. The coevolutionarily stable community is defined as the set of coevolutionarily stable equilibrium traits $(\hat{X}_1..\hat{X}_S)$ together with the set of equilibrium population sizes corresponding to those traits, $(\hat{N}_1(\hat{X}_1..\hat{X}_S)..\hat{N}_S(\hat{X}_1..\hat{X}_S))$.

In addition to the definitions above, there are additional constraints which the coevolutionary strategy model must satisfy. These constraints are that the equilibrium community, $\hat{N}_1(\hat{X}_1..\hat{X}_s)$, potentially represents a stable ecological community for all values of the traits under consideration. This constraint translates into requiring that the matrix $\partial W_i/\partial N_j$ evaluated at $N_i = \hat{N}_i(X_1..X_2)$ has eigenvalues with negative real parts for all values of the traits under consideration. This is a technical requirement which is satisfied if the ecological model within which the coevolution is embedded has these properties to begin with.

The coevolutionary strategy model, first introduced in ecology by Levins (1974, 1975), can be represented by a graph. There are nodes for the equilibrium population sizes, $\hat{N}_i$, and also for the equilibrium values of the traits $\hat{X}_i$. A coevolutionary strategy graph for two coevolving species is illustrated in Figure 1.

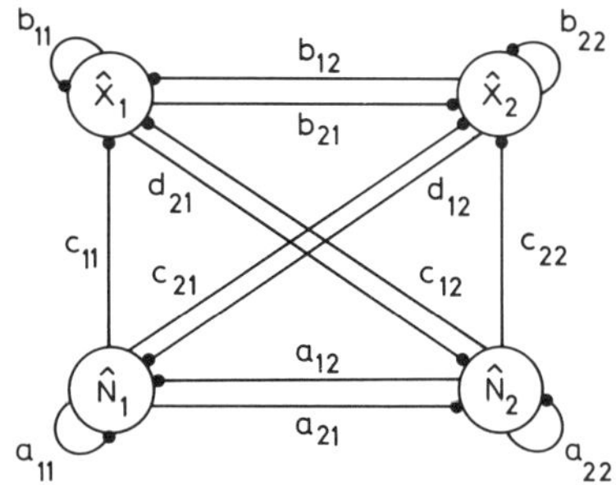

Fig. 1. The general form of a strategy graph. The variables in the model are each represented by a node, and the lines among the nodes indicate the biological connections among all the variables in the system. A line terminating in a dot indicates a negative effect by the variable from which the line originates upon the variable at which the line terminates; a line terminating in an arrow indicates a positive effect. There is a numerical weight attached to a line which describes the strength of the interaction. These weights are calculated from equations 10, 11, 12, and 13.

We define the weights attached to each connection between the $\hat{N}_i$ as

$$a_{ij} = \frac{\partial W_i}{\partial N_j} \tag{10}$$

The element a_{ij} refers to the direct effect of N_j upon the fitness in species-i. These a_{ij} must satisfy the conditions which ensure that the eigenvalues of the matrix $\{a_{ij}\}$ have negative real parts as mentioned in the paragraph above.

The connections from the X's to the N's are defined as

$$d_{ij} = \frac{\partial W_i}{\partial X_j} \tag{11}$$

The element d_{ij} refers to the direct effect of X_j upon the fitness in species-i. An important feature of the graph is that d_{ii} always equals zero in a coevolutionarily stable community. This fact is true because the $\hat{X}_i$ are defined as those values of $\hat{X}_i$ which maximize W_i and, therefore, $\partial W_i/\partial X_i = 0$ at $X_i = \hat{X}_i$.

The connections among the X's are defined as

$$b_{ij} = \frac{\partial}{\partial X_j}\left(\frac{\partial W_i}{\partial X_i}\right) \tag{12}$$

The element b_{ij} refers to the direct effect of X_j upon the slope of W_i with respect to X_i. This quantity is very important as illustrated in Figure 2.

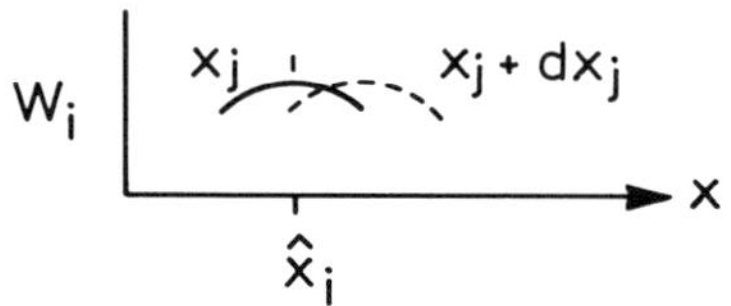

Fig. 2. How the formula for b_{ij} can be interpreted as representing the effect of x_j on $\hat{x}_i$.

The fact that W_i is maximized at $\hat{X}_i$ means that $\partial W_i/\partial X_i = 0$ at $X_i = \hat{X}_i$. If this slope becomes positive as X_j is increased to $X_j + dX_j$ then it means that the value of X_i which maximizes W_i is moved to the right. This is illustrated above. Thus, if b_{ij} is positive, it means that the direct effect of increasing X_j is to increase X_i. For this reason, the b_{ij} describes the connections between the X's. Note that b_{ii} is necessarily negative because $\hat{X}_i$ maximizes W_i.

The connections from the N's to the X's are similarly defined as

$$c_{ij} = \frac{\partial}{\partial N_j}\left(\frac{\partial W_i}{\partial X_i}\right) \tag{13}$$

The element c_{ii} refers to the direct effect of N_j on the slope of W_i with respect to X_i.

There are measures of the feedback in this graph and in various subgraphs. We begin with the subgraphs familiar from the study of pure density-dependent coevolution (Roughgarden 1977), as summarized in Table 1.

Table 1. Subgraphs and feedback measures used if evolutionary reciprocation is not being considered.

	Graph	Feedback Measure
G	graph obtained by deleting all the X's	$F = (-1)^{S+1}\det\{a_{ij}\}$
G_i	subgraph to G obtained by deleting N_i	$F_i = (-1)^{S}\det_{ii}\{a_{ij}\}$
G_{ij}	subgraph of G representing the connecting web from species-j to species-i	$F_{ij} = (-1)^{S+1}\det_{ji}\{a_{ij}\}$

In Table 1 $\{a_{ij}\}$ means the matrix of dimension S x S formed by the elements $\partial W_i/\partial N_j$. The symbol $\det_{ji}\{a_{ij}\}$ means the determinant of the matrix of dimension (S-1) x (S-1) obtained by deleting the ith column and the jth row from $\{a_{ij}\}$. By the assumption of ecological stability $F < 0$ and $\sum F_i < 0$.

The full strategy graph corresponds to the matrix of dimension 2S x 2S given by

$$Z = \begin{pmatrix} B & C \\ D & A \end{pmatrix} \tag{14}$$

The feedback to various subgraphs of the full strategy graph are defined in terms of Z matrix above. The matrix Z^{ii} is defined as the matrix of dimension (2S-1) x (2S-1) obtained by striking out the ith row and column from Z.

In Table 2, the symbol $\det_{X_j,N_k}(Z^{ii})$ means the determinant of the matrix of dimension (S-2) x (S-2) obtained by deleting the column corresponding to N_k and the row corresponding to X_j from the matrix Z^{ii}. In calculating the feedback of a connecting web, the practical points to remember are to delete the column of receiving entity and the row of the sending entity, and that, if the entities are of the same type (i.e., if both are N's or both X's), then the coefficient is $(-1)^{k+j}$; whereas, if the entities are of different types, then the coefficient is $(-1)^{1+k+j}$. These definitions of feedback follow the same idea as those introduced previously and are set out in Table 2 simply for completeness.

We will use these definitions in expressions for the effects of changing a trait in a species upon the abundances of all the species. For example, suppose that $\hat{X}_i$ is the value of the trait in species-i which maximizes W_i. Then we want to know the quantity $\partial\hat{N}_i/\partial X_i$ evaluated at this point. If this derivative is positive, it means

Table 2. Additional subgraphs and feedback measures required when evolutionary reciprocation is considered.

Graph		Feedback Measure	
H	The full coevolutionary strategy graph	I	$= -\det(Z)$
$H^{(X_i)}$	The subgraph of H obtained by deleting X_i	$I^{(X_i)}$	$= \det(Z^{ii})$
$H_{N_j}^{(X_i)}$	The subgraph of $H^{(X_i)}$ obtained by deleting N_j	$I_{N_j}^{(X_i)}$	$= \det_{N_j,N_j}(Z^{ii})$
$H_{X_j}^{(X_i)}$	The subgraph of $H^{(X_i)}$ obtained by deleting X_j $(j \neq i)$	$I_{X_j}^{(X_i)}$	$= \det_{X_j,X_j}(Z^{ii})$
$H_{N_k, X_j}^{(X_i)}$	The subgraph of $H^{(X_i)}$ representing the connecting web from X_j to N_k $(j \neq i)$.	$I_{N_k,X_j}^{(X_i)}$	$= (-1)^{1+k+j}\det_{X_j,N_k}(Z^{ii})$
$H_{N_k,N_j}^{(X_i)}$	The subgraph of $H^{(X_i)}$ representing the connecting web from N_j to N_k $(j \neq k)$	$I_{N_k,N_j}^{(X_i)}$	$= (-1)^{k+j}\det_{N_j,N_k}(Z^{ii})$
$H_{X_k,X_j}^{(X_i)}$	The subgraph of $H^{(X_i)}$ representing the connecting web from X_j to X_k $(j \neq 1 \neq i)$	$I_{X_k,X_j}^{(X_i)}$	$= (-1)^{k+j}\det_{X_j,X_k}(Z^{ii})$
$H_{X_k,N_j}^{(X_i)}$	The subgraph of $H^{(X_i)}$ representing the connecting web for N_j to X_k $(k \neq i)$	$I_{X_k,N_j}^{(X_i)}$	$= (-1)^{1+k+j}\det_{N_j,X_k}(Z^{ii})$

that if X_i were increased beyond $\hat{X}_i$, then the direct effect is to increase the abundance of the species involved. Furthermore, we want to know the quantity, $d\hat{N}_i/dX_i$, which represents the total effect on $\hat{N}_i$ of changing X_i. If this quantity is positive, then raising X_i above $\hat{X}_i$ will raise the abundance of species-i even after all the evolutionary reciprocation is taken into account. In a coevolutionarily stable community under pure density-dependent coevolution, the quantities $\partial\hat{N}_i/\partial X_i$ and $d\hat{N}_i/dX_i$ are zero. With pure density dependent coevolution, the final result is either the best or the worst for the species involved, depending on F_i (see Roughgarden 1977). But with interspecific frequency dependence the final result is somewhere in between the best and worst, and the values of $\partial\hat{N}_i/\partial X_i$ and of $d\hat{N}_i/dX_i$ at the equilibrium help to understand exactly where that somewhere is.

I. The Direct Effects

The direct effect on the $\hat{N}$'s of varying X_i around $\hat{X}_i$ is given by the following result:

In a coevolutionarily stable community, the direct effects of varying X_i are

$$\frac{\partial \hat{N}_i}{\partial X_i} = - \sum_{k \neq i} \left(\frac{F_{ik}}{F}\right) d_{ki} \tag{15}$$

$$\frac{\partial \hat{N}_i}{\partial X_j} = - \sum_{k \neq i,j} \left(\frac{F_{ik}}{F}\right) d_{kj} + \left(\frac{F_i}{F}\right) d_{ij}$$

Obviously the quantities $\partial \hat{N}_i/\partial X_j$ are generally not equal to zero, and therefore even when selection is assumed to maximize fitness in each species, the result does not maximize (or minimize) the abundance of the species involved. This result depends solely on the interspecific frequency dependence. Without frequency dependence the d_{ij} equal zero, and the results of the section on pure density dependent coevolution are regained.

II. Evolutionary Reciprocation

Interspecific frequency dependence makes evolutionary reciprocation into an important and surprising phenomenon. It can be shown that, with pure density-dependent coevolution, the total effect of evolution within a species on its own population size is dominated by the direct effect; in a coevolutionarily stable community, the contribution from the evolutionary reciprocation is zero. Reciprocation does not counteract or overpower the direct effect of evolution within a species on its own population size. But, with interspecific frequency dependence, the situation changes dramatically. In general terms, two new facts emerge. First, the evolutionary reciprocation is of the same order of magnitude as the direct effect, and so the total outcome of evolution upon a species is never determined by the direct effect alone. Second, the direction in which the reciprocation acts is often surprising. As we learned above, the direct effect of evolution does not generally lead to the optimal value of the traits involved. This problem is compounded when reciprocation is taken into account. The direct effect of evolution within a species will not only lead to an ecologically non-optimal value of a trait but it will also cause an interacting species to produce an ecologically non-optimal value of its trait. The net effect of both species evolving non-optimal values of their traits can be very surprising and can differ for each model and situation. We now explore this issue in more detail.

The total effect of a trait upon the abundance of a species is defined as the sum of the direct effect and the evolutionary reciprocation

$$\underset{\substack{\text{total}\\\text{effect}}}{\frac{d\hat{N}_i}{dX_i}} = \underset{\substack{\text{direct}\\\text{effect}}}{\frac{\partial \hat{N}_i}{\partial X_i}} + \underset{\substack{\text{evolutionary}\\\text{reciprocation}}}{\sum_{k\neq i} \frac{\partial \hat{N}_i}{\partial X_k}\frac{d\hat{X}_k}{dX_i}} \tag{16}$$

In rare circumstances this formula is actually the most efficient way to calculate the total effect. Usually, however, the most practical way to calculate the total effect is with the formulae below using the feedback measures introduced earlier.

In a coevolutionarily stable community the total effects of varying X_i are

$$\frac{d\hat{N}_i}{dX_i} = -\sum_{j\neq i}\left(\frac{I^{(X_i)}_{N_i,X_j}}{I^{(X_i)}}\right)b_{ji} - \sum_{j\neq i}\left(\frac{I^{(X_i)}_{N_i,N_j}}{I^{(X_i)}}\right)d_{ji} \tag{17a}$$

$$\frac{d\hat{N}_k}{dX_i} = -\sum_{j\neq 1}\left(\frac{I^{(X_i)}_{N_k,X_j}}{I^{(X_i)}}\right)b_{ji} - \sum_{j\neq i,k}\left(\frac{I^{(X_i)}_{N_k,N_j}}{I^{(X_i)}}\right)d_{ji} + \left(\frac{I^{(X_i)}_{N_k}}{I^{(X_i)}}\right)d_{ki} \tag{17b}$$

$$\frac{d\hat{X}_k}{dX_i} = -\sum_{j\neq i,k}\left(\frac{I^{(X_i)}_{X_k,X_j}}{I^{(X_i)}}\right)b_{ji} + \left(\frac{I^{(X_i)}_{X_k}}{I^{(X_i)}}\right)b_{ki} - \sum_{j\neq i}\left(\frac{I^{(X_i)}_{X_k,N_j}}{I^{(X_i)}}\right)d_{ji} \tag{17c}$$

The formulae for the important case of two interacting species are summarized below

Coevolutionary Mismatch for
Two Interacting Populations

Direct Effects

$$\frac{\partial \hat{N}_1}{\partial X_1} = -\left(\frac{F_{12}}{F}\right)d_{21} \qquad \frac{\partial \hat{N}_2}{\partial X_1} = \left(\frac{F_2}{F}\right)d_{21} \tag{18}$$

Total Effects

$$\frac{d\hat{N}_1}{dX_1} = \frac{\left(I^{(X_1)}_{N_1,X_2}\right)b_{21} + \left(I^{(X_1)}_{N_1,N_2}\right)d_{21}}{-I^{(X_1)}} \tag{19a}$$

$$\frac{d\hat{N}_2}{dX_1} = \frac{\left(I^{(X_1)}_{N_2,X_2}\right)b_{21} - \left(I^{(X_1)}_{N_2}\right)d_{21}}{-I^{(X_1)}} \tag{19b}$$

$$\frac{d\hat{X}_2}{dX_1} = \frac{-\left(I_{X_2}^{(X_1)}\right) b_{21} + \left(I_{X_2,N_2}^{(X_1)}\right) d_{21}}{-\ I^{(X_1)}} \qquad (19c)$$

The effects of X_2 on $\hat{N}_1$ and $\hat{N}_2$ are given by symmetrical formulae.

We illustrate the use of these formulae with two examples. The first example is of the coevolution between predator and prey, and second example is of the coevolution of character displacement between two competing species.

Example 1. Predator-Prey Coevolution

Perhaps the most classic topic in the biology of coevolution is the coevolution of predator and prey populations to one another. We visualize that X_1 denotes the effort allocated by a prey individual to predator defense and avoidance, and that X_2 denotes the effort allocated by a predator to the search for prey and to the breakdown of the prey defenses. We have discovered in general terms from the preceding theory that the prey defense effort, X_1, and the predator search effort, X_2, will not evolve to the optimum values which maximize the population sizes of the respective species. What we now want are more specifics about the ecological non-optimality of coevolution. Do the prey evolve a level of defense which is too much in the sense that lowering the defense effort would increase their population size or the reverse? Do the predators search too hard for their prey, or the reverse? These questions can all be answered using the theoretical approach developed above.

For concreteness, consider the familiar predator-prey model with logistic growth in the prey.

$$W_1 = 1 + r - \frac{r}{K(X_1)} N_1 - a(X_1,X_2)N_2 \qquad (20)$$

$$W_2 = b(X_2)a(X_1,X_2)N_1 - d + 1$$

We assume that the predation coefficient $a(X_1,X_2)$ is a monotonic decreasing function of X_1, the prey defense effort, and a monotonic increasing function of X_2, the predator search effort. Furthermore, we assume the prey defense effort, X_1, involves some cost. Therefore, let the prey carrying capacity, $K(X_1)$, be a monotonic decreasing function of X_1. Similarly, predator searching effect, X_2, involves some cost; so we let the parameter for the efficiency of conversion of captured prey into predator reproduction, $b(X_2)$, be a monotonic decreasing function of X_2. These assumptions are summarized as,

$$\partial\ a(X_1,X_2)/\partial X_1 < 0 \tag{21}$$

$$d\ K(X_1)/dX_1 < 0$$

$$\partial\ a(X_1,X_2)/\partial X_2 > 0$$

$$d\ b(X_2)/dX < 0$$

These assumptions specify a strategy model for predator-prey coevolution in sufficient detail so that the coevolutionary strategy graph can be written by inspection, and all the lines in the graph have definite signs. The graph is given in Figure 3.

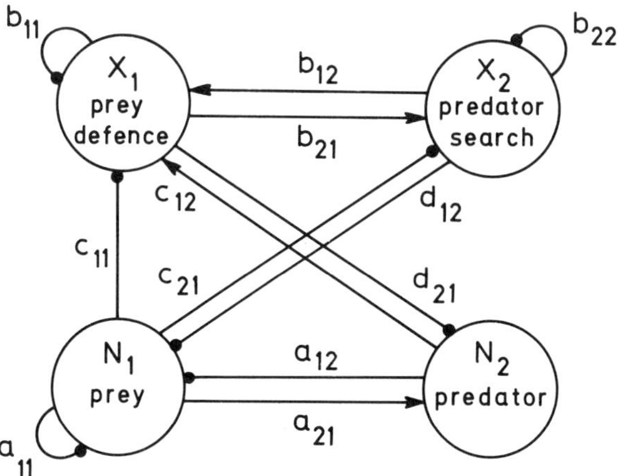

Fig. 3. Strategy graph for predator-prey coevolution based on the model in equation 20.

At the coevolutionary equilibrium, the direct effects on the population sizes of varying the traits away from their equilibrium values are given by

$$\frac{\partial \hat{N}_1}{\partial X_1} = -\ (\frac{F_{12}}{F})d_{21} = -\ \frac{d_{21}}{a_{21}} > 0 \tag{22a}$$

$$\frac{\partial \hat{N}_2}{\partial X_1} = (\frac{F_2}{F})d_{21} = \frac{a_{11}}{a_{12}a_{21}}\ d_{21} < 0 \tag{22b}$$

$$\frac{\partial \hat{N}_2}{\partial X_2} = -\ (\frac{F_{21}}{F})d_{12} = \frac{d_{12}}{a_{12}} < 0 \tag{22c}$$

$$\frac{\partial \hat{N}_1}{\partial X_2} \left(\frac{F_1}{F}\right) d_{12} \equiv 0 \qquad (22d)$$

The result for $\partial \hat{N}_1/\partial X_1$ means that increasing the level of prey defense beyond that which evolves by natural selection will actually increase the prey population size. This result does not, of course, include the possibility of evolutionary reciprocation by the predator, as discussed below. Thus, provided X_2 remains at $\hat{X}_2$, then the equilibrium level of prey defense $\hat{X}_1$ is below the optimal, and the allocation of more effort into defense at the cost of lowering K would yield net increase in the prey's abundance. The result for $\partial \hat{N}_2/\partial X_1$ is the other side of the story. If the prey were to allocate more effort into defense, then the equilibrium predator abundance would show a net decline. The result for $\partial \hat{N}_2/\partial X_2$ means that if the predators were to reduce their searching effort, X_2, below that which evolves by natural selection, then their abundance would actually increase (provided X_1 remains at $\hat{X}_1$). The result for $\partial \hat{N}_1/\partial X_2$ means that there is no net effect of varying the predator searching effect on the prey equilibrium abundance. This arises from the absence of density dependence in the prey. These direct effects may be summarized by saying that at a coevolutionary equilibrium, the prey underdefend and the predators oversearch.

The total effects of varying the traits X_1 and X_2 away from their equilibrium values include effect of evolutionary reciprocation in addition to the direct effects calculated above. First we calculate the total effect of varying X_1. The basic quantity in these calculations is $I^{(X_I)}$, the feedback in the subsystem obtained by deleting X_1. The graph and feedback are

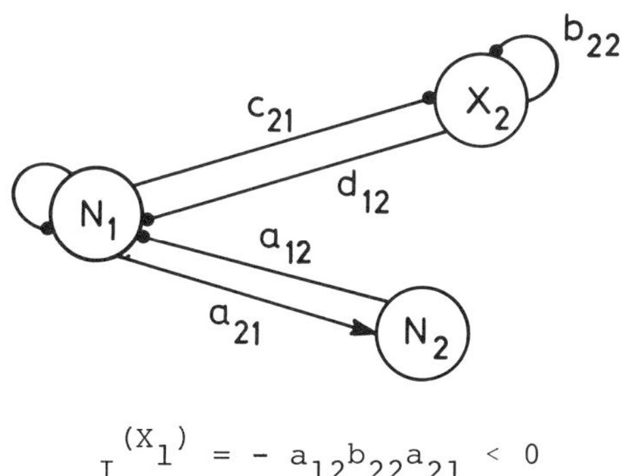

$$I^{(X_1)} = - a_{12}b_{22}a_{21} < 0$$

The feedback in this subsystem is always negative. Next we inspect various subgraphs within this subsystem. For dN_1/dX_1 we require

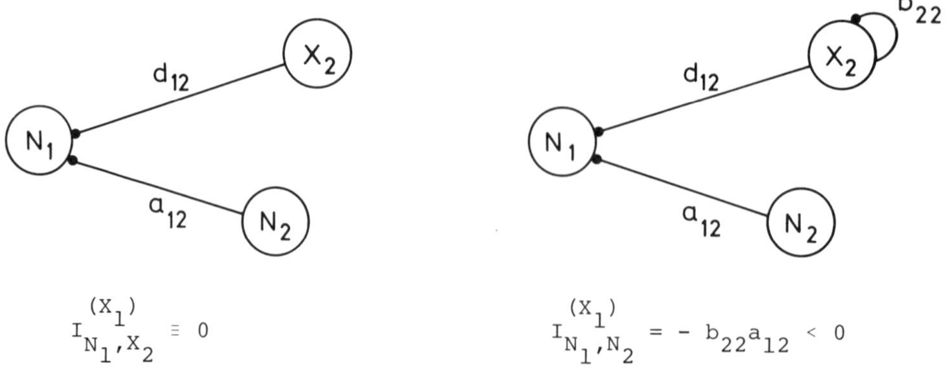

therefore

$$\frac{d\hat{N}_1}{dX_1} = \frac{(0)b_{21} + (- b_{22}a_{12})d_{21}}{- a_{12}b_{22}a_{21}} = \frac{- d_{21}}{a_{21}} > 0 \tag{23}$$

These results mean that the total effect on prey abundance of varying the level of prey defense is identical to the direct effect $\partial\hat{N}_1/\partial X_1$ as calculated earlier. Hence the effect of evolutionary reciprocation is zero in this case. This result could also be obtained by writing

$$\frac{d\hat{N}_1}{dX_1} = \frac{\partial\hat{N}_1}{\partial X_1} + \frac{\partial\hat{N}_1}{\partial X_2}\frac{d\hat{X}_2}{dX_1} \tag{24}$$

However the quantity $\partial\hat{N}_1/\partial X_2$ is identically zero as calculated above in (22d); and, therefore, the term describing the effect of evolutionary reciprocation is zero. Thus we see that even when evolutionary reciprocation is taken into account in this model, it is predicted that the prey are underdefended at the coevolutionary equilibrium.

The quantity $d\hat{N}_2/dX_1$ is obtained from the following graphs

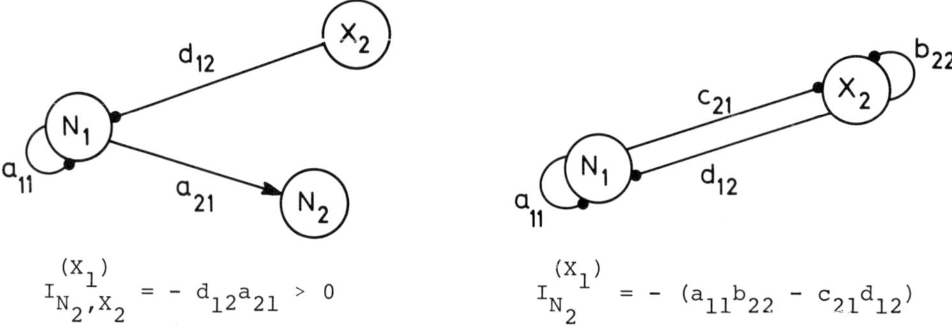

Therefore $d\hat{N}_2/dX_1$ is found to be

$$\frac{d\hat{N}_2}{dX_1} = \frac{(-\,b_{21}d_{12})}{a_{12}b_{22}} + \frac{(-\,c_{21}d_{12}d_{21})}{a_{12}a_{21}b_{22}} + \frac{a_{11}d_{21}}{a_{12}a_{21}} \qquad (25)$$

The first two terms are always positive, and the second is always negative. Therefore, the sign of $d\hat{N}_2/dX_1$ is not definitely known without more detail on the functional forms for $K(X_1)$, $a(X_1,X_2)$, and $b(X_2)$. Thus the total effect of increasing prey defense effort, X_1, on the equilibrium predator abundance involves a direct effect, which is always negative, and the evolutionary reciprocation which may or may not counteract the direct effect depending on the detailed assumptions.

The total effect of increasing prey defense on the equilibrium level of predator search effect is obtained from

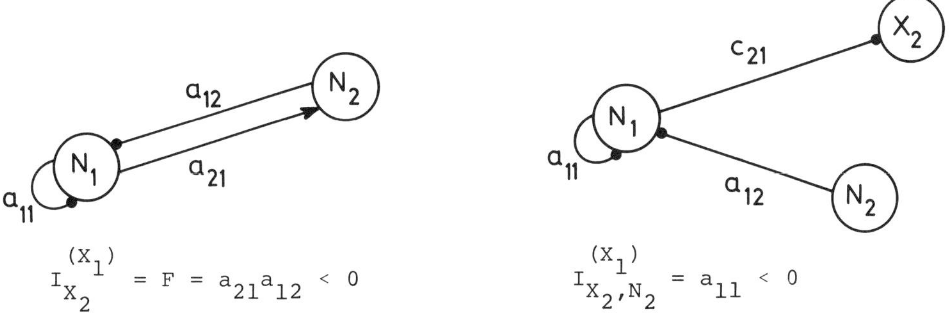

$I_{X_2}^{(X_1)} = F = a_{21}a_{12} < 0$ $\qquad$ $I_{X_2,N_2}^{(X_1)} = a_{11} < 0$

Therefore

$$\frac{d\hat{X}_2}{dX_1} = \frac{-\,a_{21}a_{12}b_{21} + a_{11}d_{21}}{a_{12}b_{22}a_{21}} > 0 \qquad (26)$$

Hence the total effect of increasing prey defense effort on predator search effort is positive.

Next consider the total effect of changing the predator search effort X_2 on the other variables in the system. The subsystem obtained by deleting X_2 is

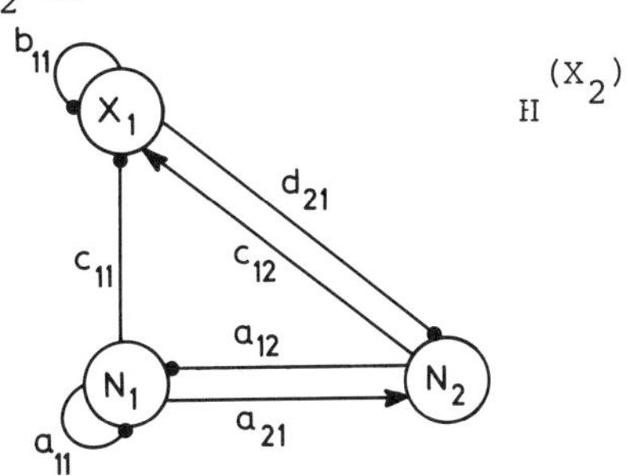

$H^{(X_2)}$

The feedback of this subsystem is

$$I^{(X_2)} = (-\ b_{11}a_{12}a_{21}) + (d_{21}c_{11}a_{12}) + (-\ d_{21}a_{11}c_{12}) < 0. \qquad (27)$$

All three terms are negative. Hence $I^{(X_2)}$ is negative.

To find the total effect of varying the search effect on the predator's equilibrium population size, we require the feedback of the following connecting webs,

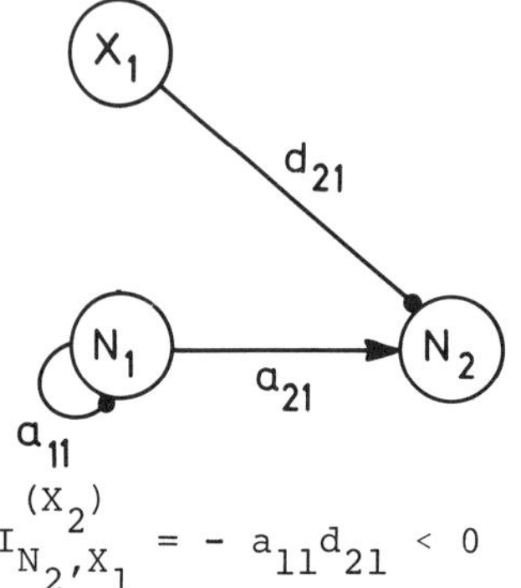

$I_{N_2,X_1}^{(X_2)} = -\ a_{11}d_{21} < 0$

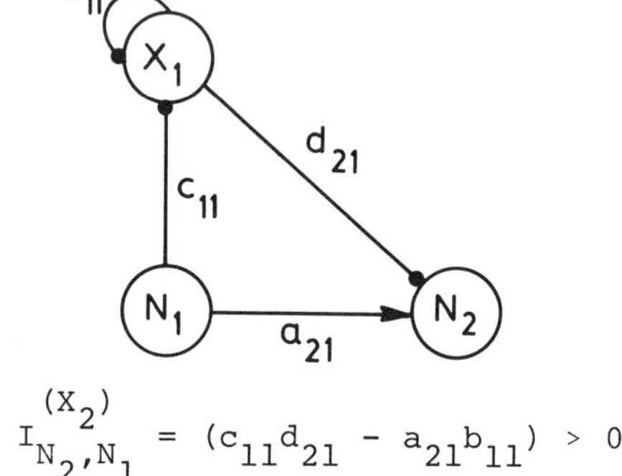

$I_{N_2,N_1}^{(X_2)} = (c_{11}d_{21} - a_{21}b_{11}) > 0$

Hence we have

$$\frac{d\hat{N}_2}{dX_2} = \frac{-\ a_{11}d_{21}b_{12} + (c_{11}d_{21} - a_{21}b_{11})d_{12}}{(-I^{(X_2)})} < 0 \qquad (28)$$

The numerator is always negative. Therefore $d\hat{N}_2/dX_2$ is negative.

Recall that the direct effect, $\partial\hat{N}_2/\partial X_2$, is also always negative. This result shows that the total effect, which includes the evolutionary reciprocation from the prey, has the same sign as the direct effect. Hence, we again predict that $\hat{X}_2$ is too high. That is, if the predators were to allocate less effort to the search for prey, then their population size would show a net increase even when evolutionary reciprocation by the prey is taken into account.

To determine the total effect of X_2 on the equilibrium prey abundance, we inspect

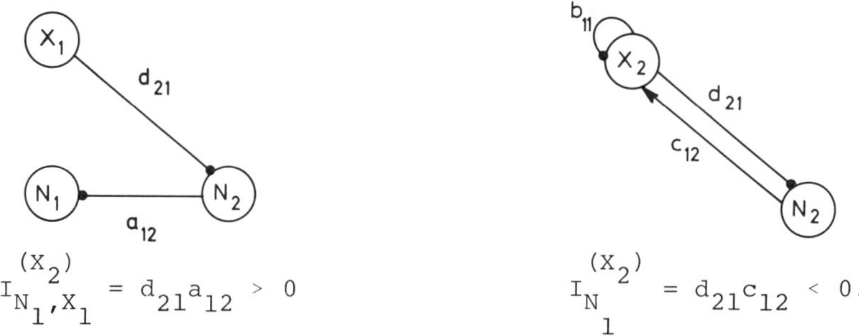

$$I_{N_1,X_1}^{(X_2)} = d_{21}a_{12} > 0 \qquad I_{N_1}^{(X_2)} = d_{21}c_{12} < 0.$$

Hence

$$\frac{d\hat{N}_1}{dX_2} = \frac{(d_{21}a_{12}b_{12}) + (-\,d_{21}c_{12}d_{12})}{(-I^{(X_2)})} \tag{29}$$

The first term in the numerator is positive, and the second is negative. Therefore the total effect of X_2 on $\hat{N}_1$ depends on the detailed shapes of functions $K(X_1)$, $a(X_1,X_2)$, $b(X_2)$. The interesting aspect of this result is that the direct effect of X_2 on $\hat{N}_1$, $\partial\hat{N}_1/\partial X_2$, is identically zero. Therefore, the total effect above is caused solely by the evolutionary reciprocation.

Similarly the total effect of varying the predator search effort on the level of defense evolved by the prey is found to be

$$\frac{d\hat{X}_1}{dX_2} = \frac{(-\,a_{12}a_{21}b_{12}) + (a_{21}c_{12}d_{12})}{-I^{(X_2)}} \tag{30}$$

The first term in the numerator is positive, and the second is negative. Thus, the total effect of varying X_2 on $\hat{X}_1$ depends on more detail than is contained in the signs of the connections in the strategy graph.

To summarize, at a coevolutionary equilibrium, (1) the prey underdefend themselves ($d\hat{N}_1/dX_1 > 0$); (2) increasing prey defense will cause increased predator searching effort to evolve ($d\hat{X}_2/dX_1 > 0$); and (3) the predators search too hard ($d\hat{N}_2/dX_2 < 0$). To determine the total effect of varying X_1 and X_2 on the equilibrium values of the other variables in the system requires information not only on the signs, but also the magnitudes of the connections in the strategy graph.

Example 2. <u>Coevolution of Character Displacement</u>

This example concerns the evolution of the competition coefficients between two competitors according to the scheme illustrated in Figure 4. In each species there is a trait which indicates the kind of resources used by that species. The carrying capacity of a species depends on the value of its trait according to the function

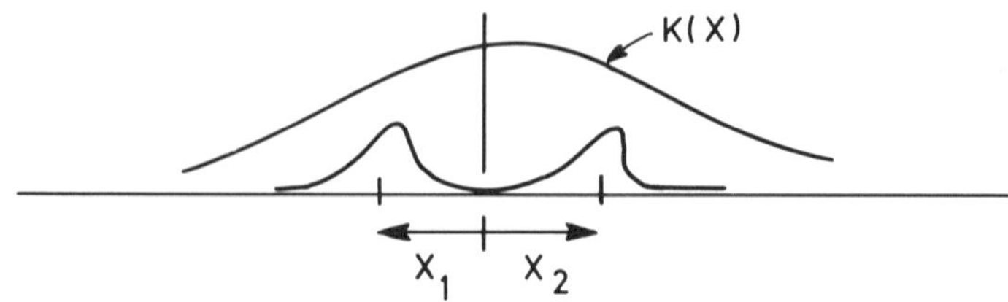

Fig. 4. A setup for the model of the coevolution of niche positions of two competing species.

K(X). The competition coefficient between the competitors is determined by the value of the traits in both species according to the function $\alpha(X_1 + X_2)$. Thus the strategy model is given by the following formulae

$$W_1 = 1 + r - \frac{r}{K(X_1)} N_1 - \frac{r\,\alpha(X_1 + X_2)}{K(X_1)} N_2$$
$$W_2 = 1 + r - \frac{r}{K(X_2)} N_2 - \frac{r\,\alpha(X_1 + X_2)}{K(X_2)} N_1 \qquad (31)$$

There is full symmetry in the model. This model leads to the strategy graph in Figure 5.

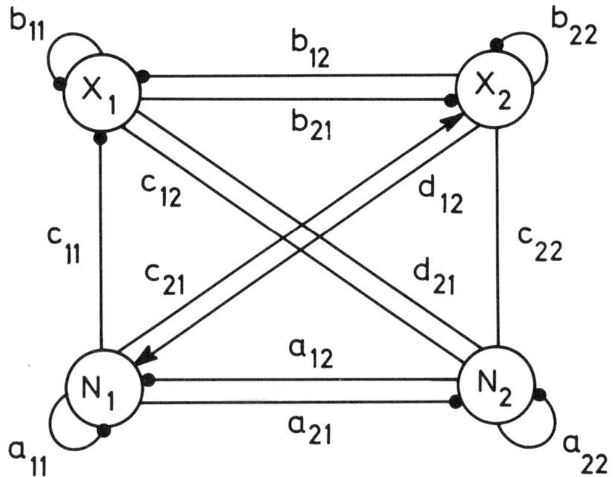

Fig. 5. Strategy graph for the coevolution of the niche positions of two competing species.

With this example, we illustrate two interesting biological points. First, the concept of evolutionary reciprocation depends upon the assumption of a coevolutionary *path*. Second, we will determine a very general fact about the mismatch in the coevolution of character displacement.

The total effect of varying a trait, say X_1, on the population size of a species, say $\hat{N}_1$, is formally determined by the chain rule in calculus

$$d\hat{N}_1 = \frac{\partial \hat{N}_i}{\partial X_1} dX_1 + \frac{\partial \hat{N}_1}{\partial X_2} (\frac{dX_2}{dX_1}) dX_1 \tag{32}$$

The quantity dX_2/dX_1 describes the path that X_2 takes as X_1 varies. In the entire preceding discussion, we have assumed that X_2 moves along the function $\hat{X}_2(X_1)$ as X_1 is varied. This is one meaningful choice of a path, and probably the most important choice for most purposes. But it is not the only possible and useful choice of a path, as we shall see.

Consider one of the possible *dynamic* paths taken by the competitors during their coevolution of character displacement. Suppose both species start out with exactly symmetrical initial positions very near the center of the K(X) function, i.e., $X_1(0) = X_2(0) = \varepsilon$. Then as time proceeds both species will move away from one another until they stop at the coevolutionary equilibrium point. Along this path, each step moved by species-1 is matched by an equal step in the opposite direction by species-2, so that

$$dX_1/dX_2 \equiv dX_2/dX_1 \equiv 1 \tag{33}$$

The total change in the equilibrium population size of species-1 along this path then reduces to

$$d\hat{N}_1 = (\frac{\partial \hat{N}_1}{\partial X_1} + \frac{\partial \hat{N}_1}{\partial X_2}) dX_1 \tag{34}$$

At the coevolutionary equilibrium, the direct effects are known in terms of the feedbacks, so that

$$\frac{d\hat{N}_1}{dX_1} = \frac{-F_{12}}{F} d_{12} + \frac{F_1}{F} d_{12} \tag{35}$$

This expression gives the direct effect of varying X_1 on $\hat{N}_1$ based on assuming that X_2 shifts in a symmetric way. If $d\hat{N}_1/dX_1$ is positive at the coevolutionary equilibrium, it means that both species have not moved far enough apart. If they both moved farther apart, they would then realize a net increase in population size.

To determine the sign of $d\hat{N}_1/dX_1$ we evaluate the feedbacks as

$$F_{12} = a_{12} = \frac{\partial W_1}{\partial N_2} = \frac{-r\,\alpha(X_1 + X_2)}{K(X_1)} \tag{36a}$$

$$F = a_{11} = \frac{\partial W_1}{\partial N_1} = \frac{-r}{K(X_1)} \tag{36b}$$

So we obtain

$$\frac{d\hat{N}_1}{dX_1} = \frac{-r}{K(X_1)}\left[1 - \alpha(X_1 + X_2)\right]\frac{d_{12}}{F} > 0 \qquad (37)$$

Since F is negative, we have that $d\hat{N}_1/dX_1$ is positive, provided $1 > \alpha(X_1 + X_2)$ which is always true. Thus, without specifying the functional form of the carrying capacity and competition functions, apart from symmetry, we conclude generally that the process of character displacement between two symmetrical competing species does not proceed far enough. A very special case of this result was derived in Roughgarden (1976) based on a Gaussian K(X) and α(X).

The fact that two symmetrical coevolving species never displace far enough apart from one another is all the more surprising when contrasted with the evolutionary result for a single species evolving with respect to an evolutionarily rigid competitor. Suppose X_2 has some fixed value and that species-1 is evolving. At the evolutionary equilibrium for species-1 we have

$$\frac{d\hat{N}_1}{dX_1} \equiv \frac{\partial\hat{N}_1}{\partial X_1} = \frac{-F_{12}}{F}\, d_{12} < 0$$

Thus a single species displacing with respect to an evolutionarily rigid competitor always displaces too far.

References

Levin, S.A., Udovic, J.D.: A mathematical model of coevolving populations. Amer. Natur. 111:657-675. (1977).

Levins, R.: Qualitative analysis of partially specified systems. Ann. N.Y. Acad. Sci. 231:123-138. (1974).

Levins, R.: Evolution in communities near equilibrium. In Ecology and Evolution of Communities. (ed. M.C. Cody and J.M. Diamond). Harvard: Belknap, 1975, pp. 16-50.

Maynard Smith, J.: The theory of games and the evolution of animal conflict. J. Theoret. Biol. 47:209-221. (1974).

Roughgarden, J.: Resource partitioning among competing species - a coevolutionary approach. Theor. Pop. Biol. 9:388-424. (1976).

Roughgarden, J.: Coevolution in ecological systems II. Results from "loop analysis" for purely density-dependent coevolution. In Symposium on Measuring Selection in Natural Populations. (ed. T. Fenchel and F. Christiansen). New York: Springer-Verlag, 1977.

2 Physiology, Biochemistry, and Adaptation

Physiology and Biochemistry of Enzyme Variation: The Interface of Ecology and Population Genetics

R.K. Koehn

A. Introduction

"That is to say, the genes in question, insignificant as are their visible effects, have an important influence on the physiology of the organism, modifying profoundly the individual as a unit upon which selection operates." E.B. Ford, 1965.

Since ecological genetics has constituted a unified subdiscipline of evolutionary biology for several decades, it might seem somewhat unnecessary now to be focusing on the interface of ecology and genetics. Ecological genetics is concerned with the study of adaptation, how the genetic constitution of a population behaves during the course of the population's evolution, and the identification of the ecological factors that initiate and direct this evolution. Since it is at the interface of genetics and ecology where evolutionary forces are focused, it is here that we must explore the workings of the evolutionary process.

During the past decade, ecological genetics has come to be dominated by efforts to understand the interdependence of phenotypic diversity, genotypic variation, and ecological variables in the maintenance of enzyme polymorphism. Recent major treatments of the problem have been presented by Lewontin (1974), Nei (1975), and papers in Ayala (1976). It is now widely accepted that genetic variation at loci encoding for soluble enzymes is ubiquitous in natural populations of most plant and animal species. Allele frequencies and their heterozygosites are often spatially or temporally correlated with variation of ecological factors (Bryant, 1974; Hedrick et al., 1976; Koehn et al., 1976; and others). These observations, while intuitively compelling, mainly represent inferential data that enzyme polymorphism is adaptive (Lewontin, 1974). The point that such studies approach, but do not directly answer, is whether or not natural selection is directed at the enzyme loci under examination. Various ecological factors may differentially influence individual phenotypes when phenotypic diversity exists. This will result in a deterministic change in the underlying constellation of genotypes. Unfortunately, there is a dearth of demonstrable phenotypic diversity among enzyme genotypes. A successful demonstration of the adaptive significance of genetic variation of enzymes is dependent not only upon the identification of phenotypic diversity, but also on evidence that the discovered phenotypic diversity is ecologically relevant. Since we are here investigating molecular variation, it is in the properties of molecules that we must seek phenotypic diversity. In the study of enzyme polymorphism the interface of ecology and genetics lies squarely in the fields of biochemistry and physiology. It is this point which I hope to emphasize in the following presentation, first by delineating the criteria necessary for the description of evolutionarily important enzyme polymorphism and second, by a detailed example of how these criteria may be fulfilled from a study of leucine aminopeptidase variation in the marine intertidal bivalve, *Mytilus edulis*.

B. When is Enzyme Polymorphism Adaptive?

It is important to consider the kinds of information that we would require in order to establish unequivocally the adaptive importance of enzyme variation. This is especially true as increasing numbers of investigators turn to the description of biochemical differences among enzyme phenotypes in order to establish the evolutionary relevance of individual polymorphisms.

1. Phenotypic diversity must exist among genotypes in some measure of molecular function. Detailed biochemical studies of individual enzyme polymorphisms have demonstrated functional differences among allozyme variants (Koehn, 1969; Merritt, 1972; Gibson, 1970; Vigue and Johnson, 1973; Day et al., 1974a, b; Day and Needham, 1974; Thorig et al., 1975; and Miller et al., 1975). Such studies have been cited as important evidence that enzyme polymorphism is adaptive. Described differences often emerge in different enzymological parameters in each investigation, including maximum velocity (V_{max}, or "activity"), pH optima, Michaelis constant (K_M), and so forth. These investigations have been collectively directed toward a structurally and functionally diverse set of enzymes, within different species; as yet particular kinetic properties have not emerged as more important than others. There is often an intuitive connection between the nature of the discovered molecular variation and the distribution of individual phenotypes in natural populations. Such a correlation, while heuristically pleasing to the "selectionist" as an argument that the polymorphism is adaptive, is nevertheless inferential. Still, differences in molecular function must be described, as natural selection cannot discriminate among genotypes unless at least some individual genotypes are phenotypically unique. While the existence of such differences is a necessary condition, it is not sufficient, since it must be established whether or not the described functional differences have different physiological manifestations. Functional differences among phenotypes, like correlations between environmental variables and allele frequency variations in nature, can help pinpoint those ecological factors that challenge an organism's adaptive ability.

2. A specific locus must be affected by a specific ecological component, via the enzyme product of the locus. We must be able to demonstrate that when some ecological component varies it has an effect on the biochemical reactivity or physiological context in which the enzyme functions. This is not to say that a locus might be affected by other ecological components as well, or that other loci may be affected by the same ecological components. It is important, however, to demonstrate the direct influence of a specific ecological variable upon the function of an individual enzyme.

3. Described phenotypic (functional) diversity must be physiologically relevant. While phenotypic diversity of enzyme function could represent an adaptation to the variation in some ecological component, it is the *differential* effects of allozyme function that must be manifest in the physiological and biochemical context within which the enzyme functions. Certainly the concept of enzyme polymorphism as an adaptation to a heterogeneous environment is not new (Levene, 1953; Levins, 1968; empirical studies reviewed by Hedrick et al.,

1976). Nevertheless, describing functional differences among allozymes whose frequencies are *correlated* with environmental gradients or patches is not enough; it still needs to be shown that the described functional differences have different effects on an organism's ability to adapt to these variations. We may think of ecological variation as a property of the cellular milieu, or as a variable in the physical environment. In either case, when such variations confront an organism, the different molecular properties of the individual allozymes must serve as a mechanism to maintain some optimality of function. This optimality may be in reaction rates, biochemical efficiency, degree of modulation, or some other biochemical aspect of maintaining a homeostatic physiology in the face of ecological perturbations.

4. Differences among phenotypes must be ultimately manifested as differences in some measure of fitness. Differences in genotypic fitness are a natural consequence of physiologically significant biochemical variation among phenotypes. Establishing the relationships above is in fact a description of the biochemical and physiological mechanisms that might underlie differences in viability, fecundity, fertility, and so forth.

Knowledge of the foregoing interrelationships among functional variation, physiological manifestation, and environmental variation is necessary to demonstrate unequivocally the adaptive significance of an enzyme polymorphism. In specific cases, other critical interrelationships might need to be established, while in still others, not all of the foregoing points would be relevant. For example, certain genotypes may be most adaptive in any ecological situation (Somero and Low, 1977) in which case it would not be possible to "map" phenotypes onto environments. Although the foregoing discussion has been in terms of a single enzyme locus, these same basic points would be addressed in, for example, comparisons of enzymes of related function or comparisons of specific metabolic pathways.

Correlation analyses between ecological and genetic variables have been a popular means of inferentially demonstrating the adaptive significance of enzyme polymorphism. When such correlations are observed, it is implicit that the relationships described above exist as the underlying causative mechanism. This is true only if the correlation reflects cause and effect between the correlated variables.

I have described what I feel to be the minimal information necessary to establish the adaptive significance of an enzyme polymorphism. The study described below is an attempt to gather this information and to exemplify how the general problem of enzyme variation might be experimentally approached. To these ends, the study of the leucine aminopeptidase polymorphism in *M. edulis* is informative, since prior to these studies, nothing was known of its biochemical properties or physiological function.

C. Spatial variation of leucine aminopeptidase: definition of the problem

Genetic variation at the leucine aminopeptidase (*Lap*) locus is phylogenetically widespread in marine bivalves (Koehn and Mitton,

1972; Levinton and Koehn, 1976). Five allelic forms can be observed in populations of *Mytilus edulis*, though only three are common. These have been designated *Lap*^{98}, *Lap*^{96} and *Lap*^{94}, in order of decreasing electrophoretic mobility of their products (Koehn et al., 1976). During the past few years, several observations have been made that suggest the polymorphism to be adaptive. There are microgeographic differences throughout the tide zone in the frequencies of Lap phenotypes as well as certain dilocus phenotypic combinations between the *Lap* locus and another aminopeptidase locus (Mitton and Koehn, 1973). There are also incongruous spatial patterns of variation of allele frequencies among several loci in *Mytilus edulis* throughout the Atlantic coast of North America (Koehn et al., 1976). When different patterns of variation are observed at individual loci, selection must be contributing to at least some. Since spatial differentiation is greatest at the *Lap* locus, this locus seems most likely to be under the influences of natural selection.

There is covariation of *Lap* phenotypic frequencies between *Mytilus edulis* and another mussel, *Modiolus demissus*, which shares alleles with it, over a spatial gradient in the environment (Koehn and Mitton, 1972). Apparent viability differences exist between at least the *Lap*^{94} allele and the other *Lap* alleles (Milkman and Koehn, 1977). Finally, there is striking environmental-genotype correlation between the frequency of *Lap* alleles and spatial variations in environmental salinity and temperature (Koehn et al., 1976). The frequencies of the *Lap* alleles vary among populations of *M. edulis* on the east coast of North America. In all nonestuarine samples from Virginia to Cape Cod, the frequency of *Lap*^{94} is generally never lower than 0.48 nor higher than 0.60, with an average value of 0.56. North of Cape Cod the frequency of this allele abruptly decreases to approximately 0.10; and intermediate frequencies can be observed in the Cape Cod Canal, which connects these two major geographic regions. The frequency of *Lap*^{94} is 0.10 throughout the Gulf of Maine. In geographic regions south of Cape Cod, there are significant differences in the relative allele frequencies between coastal and estuarine populations. This difference is particularly dramatic among samples taken at the entrance of Long Island Sound from the Atlantic Ocean, a region of reduced salinity and other estuarine characteristics (Figures 1 and 2).

At this point, it is important to emphasize that these data on *Lap* variation in *Mytilus edulis* are not unique, but consist of a description of a pattern in nature and inferences, *vis à vis* selection, drawn from it. The essence of this problem, as well as many other studies of a similar nature, is illustrated in Figure 2. A particular pattern of spatial variation in genetic composition has been observed, in this case a very abrupt step cline in the frequency of *Lap*^{94}. The pattern of spatial variation is correlated with a change in an environmental component, namely salinity. In view of the extended larval dispersal that occurs in this species, a period between approximately 21 and 35 days, the marked pattern of spatial heterogeneity is unexpected. It was on the basis of this unexpected observation that Koehn et al. (1976) concluded that such differentiation must be due to differential effects of change in salinity on the individual phenotypes at this locus. The same rationale has been used by countless other investigators in studies of other enzyme polymorphisms. Nevertheless, the conclusion that natural selection is operating in the maintenance of this spatial pattern is still an inferential one. There is no evidence that natural selection, if it is

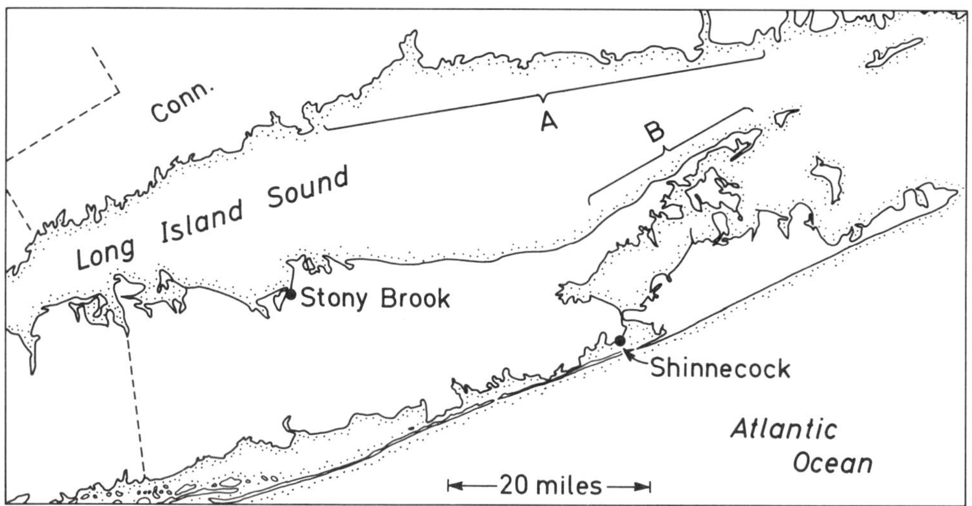

Fig. 1. The geographic location of two study populations and the geographic distribution of clinal variation of Lap^{94} illustrated in Fig. 2(A) and Fig. 7(B).

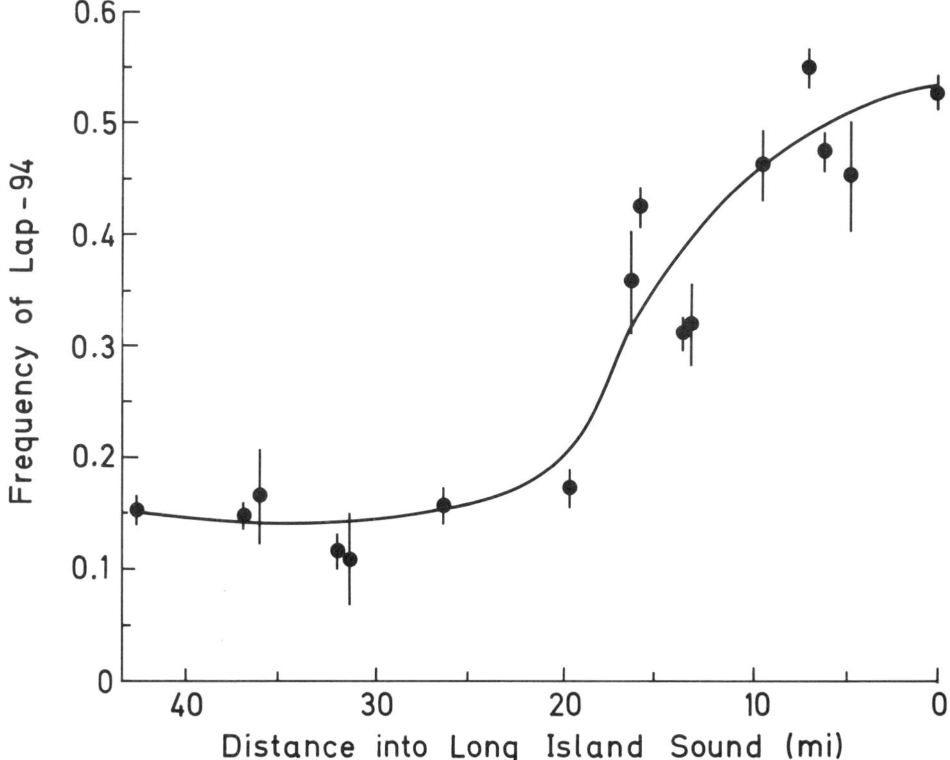

Fig. 2. Clinal variation in the frequency of the Lap^{94} allele at the eastern entrance to Long Island Sound on the Connecticut shore. Data from Lassen and Turano (unpubl.).

indeed important, is directed at the *Lap* locus, as opposed to some other chromosomally associated locus or loci. However, I will now begin to implicate directly the action of natural selection at the *Lap* locus, using information on biochemical and physiological function.

D. Functional diversity among Lap phenotypes

There is significant phenotype-dependent leucine aminopeptidase activity in samples taken from the Stony Brook and Shinnecock study sites (Table 1; Figure 1). In both populations, individuals possessing the Lap^{94} allele, in either heterozygous or homozygous condition, exhibit a greater average level of enzyme activity. It is the frequency of this allele which changes sharply between oceanic and estuarine habitats (Koehn et al., 1976; Fig. 2).

Of 18 samples (n $\approx$ 80 for each), taken from both populations up to August, 1976, there were seven occurrences of significant among genotype variation in average enzyme activity. Since 1976, an additional 20 smaller samples have been taken. In these 38 population samples taken during the past two years, in only eight did a genotype not involving the Lap^{94} allele rank highest in average enzyme activity. We may conclude that there are significant phenotype-dependent differences in average *Lap* activity. These differences may be due to either functional variations of the individual allozymes composing these phenotypes or differences in the average concentration of enzyme protein among the six phenotypes.

The leucine aminopeptidase enzyme is strongly inhibited by the presence of divalent cations, especially those of heavy metals. This is a characteristic property of thiol peptidases (Barrett and Dingle, 1971). Partial inhibition occurs in the presence of 0.01 mM cations while total inhibition is effected by 1.0 mM (Koehn, 1978). Two concentrations (0.04 and 0.01 mM) of zinc (as chloride) giving partial overall inhibition of enzyme activity were used to estimate the relative inhibition of *Lap* phenotypes.

The relative inhibition by both zinc concentrations was different in samples from the two study populations, ranging between 64 to 47 percent of controls in Stony Brook animals and between 83 and 68 percent of controls in Shinnecock animals for .04 nM Zn^{++} (Table 2). The different responses between populations are probably related to the different levels of enzyme activity in these samples (see below) since the experiment was carried out with crude homogenates. There is a large variation among individual *Lap* phenotypes in their relative sensitivities to the presence of zinc. In general, phenotypes of the Lap^{94} allele were inhibited more than other phenotypes, though the inhibition of *Lap*-96/96 was also high (Table 2). The phenotypic dependence of the inhibition is extraordinarily strong; more than 70 percent of the variance in percent remaining activity can be attributed to phenotype differences.

These observations, a significant phenotypic dependence of enzyme activity in natural populations and differences among *Lap* phenotypes in sensitivity to zinc, do not demonstrate, nor even provide a clue to, the possible adaptive significance of the functional diversity. If enzyme activity *per se* could be shown to be of some relevance, these observations might be taken as inferential evidence that the

Table 1. Phenotype-dependent leucine aminopeptidase activity in the Stony Brook (low salinity) and Shinnecock (normal salinity) populations. Enzyme activity is expressed in activity units, where 1 unit equals an E_{510} of .001, or the liberation from L-leucyl-β-naphthylamide of 6 μ moles of β-naphthylamine per minute at 37°C and pH 7.0 (.2 M Tris-HCl). Adapted from Koehn, 1978.

	Lap-98/98	-98/96	-96/96	-98/94	-96/94	-94/94	Total	F^a
Stony Brook								
Mean	208,466	247,943	271,304	278,125	329,500	333,042	249,800	5.79^c
N	22	28	14	6	2	4	76	
Shinnecock								
Mean	266,903	356,405	320,833	260,467	378,042	366,594	355,011	5.62^c
N	6	7	4	27	8	28	80	
t Statistic	1.942	5.912^c	1.200	5.876^c	---	1.188	13.140^b	
degrees of freedom	26	33	16	31	---	30	10	
P	> .05	< .001	> .20	< .001	---	> .20		

[a] One-Way Analysis of Variance

[b] $P < .005$

[c] $P < .001$

<u>Table 2.</u> Phenotype-dependent inhibition of leucine aminopeptidase activity by .01 and .04 mM zinc (as chloride). Enzyme activities are expressed as in Table 1, but as percent of controls.

	mM Zn^{++}	N	Phenotype						F^a
			98/98	98/96	96/96	98/94	96/94	94/94	
Stony Brook									
	.04	74	64	60	47	47	51	--[d]	21.6^c
	.10	74	33	36	22	21	28	--	25.7^c
Shinnecock									
	.04	75	83	81	68	73	74	74	2.9^b
	.10	75	17	19	13	15	15	17	2.6^b

[a]One-Way Analysis of Variance;

[b]$P < .05$,

[c]$P < .001$

[d]No individuals in sample

polymorphism is adaptive, though some underlying mechanism need be described. For this, we must elucidate some known biochemical and physiological functions of the leucine aminopeptidase enzyme wherein activity variations might be important. The differences among the phenotypes do, however, demonstrate that functional diversity exists among the individual Lap phenotypes. They are not enzymologically equivalent, and this could provide a target upon which natural selection acts. In order to deduce whether these differences are adaptive, we must explore in greater detail the biochemistry of the leucine aminopeptidase enzyme as well as its role in the physiology of this organism.

E. Physiological and Biochemical Function of Leucine Aminopeptidase

The leucine aminopeptidase enzyme from *Mytilus edulis* has been taken to near absolute purity, and its biochemical properties have been explored in detail (Young et al., 1978). A summary of the substrate affinities of Lap is presented in Table 3. Although a great diversity of aminoacyl naphthylamides and/or oligopeptides may serve as substrate, Lap has highest affinity for those which possess a neutral or aromatic amino group. Hence, naphthylamide derivatives of tryptophan, glycine and leucine are excellent substrates, as are oligopeptides when these residues are in the N-terminal position. While aminoacyl naphthylamides are synthetic compounds, oligopeptides are ubiquitously distributed in living cells.

Table 3. A summary of the peptidase properties of leucine aminopeptidase enzyme from *Mytilus edulis* (from Young et al., in prep.).

SUMMARY OF PEPTIDASE ACTIVITY

1. Cleaves neutral and aromatic amino acyl naphthylamides
2. Cleaves dipeptides of neutral and aromatic residues
3. No activity with NH_2-blocked dipeptides or NAP derivatives
4. Normal activity with COOH-blocked dipeptides
5. No proteinase activity.

The *Lap* enzyme has no affinity for NH_2-blocked dipeptides or naphthylamide derivatives but exhibits normal activity with COOH-blocked dipeptides as substrate. There is no evidence that Lap has an affinity for cleaving native proteins in any other but the N-terminal position.

The salient point of these biochemical characteristics is that when the N-terminal residue is either neutral or aromatic, the aminopeptide will be cleaved by the leucine aminopeptidase enzyme, thus liberating a free amino acid. Hence, we must look towards some aspects of the protein metabolism of *Mytilus edulis* for the physiological role of the leucine aminopeptidase enzyme.

Mammalian lysosomal enzymes can serve as a useful model with which to compare the potential function of leucine aminopeptidase in *M. edulis*. Lysosomal peptidases in mammals apparently function, in concert with lysosomal proteinases, in the catabolism of cellular constituents and serve to maintain normal cellular economy (Barrett and Dingle, 1971). In addition to the normal functions of protein turnover, excretion, reabsorption, and so forth, there is increased activity in lysosomal hydrolases during starvation in mammals with a consequent mobilization of cellular protein energy reserves (Kim et al., 1973). Presumably, these same cellular functions occur in *Mytilus edulis*, but with the important addition of one other physiological function. The cells of marine bivalves, including *Mytilus*, remain isoosmotic with environmental salinity by changes in the concentration of free amino acids (Bricteux-Grégoire et al., 1964). When organisms are exposed to increases in salinity, the concentration of the amino acids, or the size of the free amino acid pool, is elevated. Whether the increase in amino acids is due to *de novo* synthesis or catabolism of cellular proteins is not known (Bayne et al., 1976), although a protein source is presently favored (Bishop, 1976). When environmental salinity declines, there is active excretion of the free amino acids, together with deamination and release of ammonia, resulting in a reduced concentration of cellular amino acids. Leucine aminopeptidase activity has a lysosomal subcellular localization (M. Moore, personal communication), is affected by changes in environmental salinity (see below); and enzyme activity is positively correlated with salinity variation. Enzyme activity is therefore also positively associated with the concentration of cellular free amino acids. Since the reaction product of the hydrolysis of substrate by *Lap* is a free amino acid, we have performed a detailed investigation of the changes in enzyme activity and its histochemical distribution during the course of salinity acclimation.

Not unlike the situation in mammals, naphthylamidase activity can be localized in *M. edulis* to both the brush border of gut epithelial cells and digestive tubule cells in the hepatopancreas. (The hepatopancreas in *Mytilus edulis* is analogous to mammalian liver.) Digestive tubule cells, lining ramifications of the stomach and intestine, are sites of active cellular metabolism. They are not only important in the intracellular digestion of ingested food (principally carbohydrates and lipids), but also serve as a prominent site of glycogen and protein storage (Thompson et al., 1974). In mammals, naphthylamidase activity in liver tissue is due to lysosomal cathepsins, while brush border naphthylamidase activity in the gut is enzymologically distinct from the cathepsins (Maroux et al., 1973; Kim et al., 1974b; Wojnarowska and Gray, 1975). Brush border peptidases are thought to function in the active transport of the products of intraluminal protein digestion (i.e. amino acids) across the gut epithelial cell membrane (Smyth, 1972; Adibi and Mercer, 1973; Kim et al., 1974a; Piggott and Fottrell, 1975; Kim and Brophy, 1976).

In order to discover which of the two principal sites of naphthylamidase activity in *M. edulis* hepatopancreas and gut were due to the *Lap* enzyme of interest, anti-*Lap* antibodies were prepared by injection into rabbits of essentially pure enzyme. Antibodies were used with commercially obtained sheep anti-rabbit fluorescent labeled antibodies in order to localize the cellular distribution of the polymorphic Lap enzyme. When this was done, fluorescent labeling was observed in both the brush border of the gut epithelial cells *and* the digestive tubules. Hence, this enzyme is distributed in two completely different tissue localizations (Fig. 3A), a situation quite unlike that in mammals.

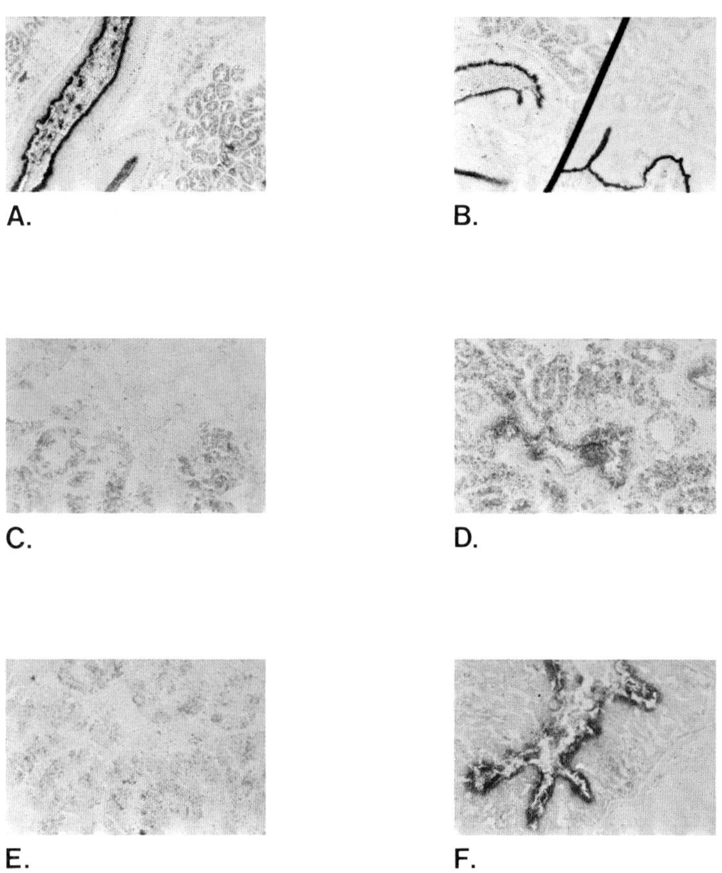

<u>Fig. 3.</u> Cytochemical localization of leucine aminopeptidase activity in the gut and digestive tubules of *M. edulis*. Enzyme activity was detected with L-leucyl-4-methoxy-β-naphthylamide and is apparent as the deposition of both red and blue complexes (depending on intensity of reaction). Yellow coloration in sections is due to a natural pigment, not *Lap* enzyme activity. A: a control individual, acclimated to 33 ppt salinity, showing digestive tubule (lower right) and intestinal brush border (left) activities; 10X magnification. B (left panel): same as A, but a field animal from Shinnecock, L.I., a normal oceanic salinity (33 ppt) population; 10X. B (right panel):

same as A, but a field animal from Stony Brook, L.I., an estuarine (≈25 ppt) population; 10X. C: low enzyme activity in the digestive tubule cells of an animal acclimated to 15 ppt salinity; 90X D: high enzyme activity in digestive tubule cells in a 15 ppt acclimated animal, 12 hours after exposure to 33 ppt salinity; 90X. E: low enzyme activity in digestive tubule cells in a 33 ppt acclimated animal, 6 hours after exposure to 15 ppt salinity; 25X. F: cellular damage to intestinal epithelial cells and fragmentation of the brush border and associated enzyme activity in a 15 ppt acclimated animal, 5 days after exposure to 33 ppt salinity.

Three cellular locations were monitored by microdensitometry during salinity acclimation experiments. These were the brush border of the gut epithelial cells, the cytoplasm of the gut epithelial cells and the digestive tubule cells. At the same time, spectrophotometric assays were performed on extracts of hepatopancreas tissue at each sampling time during an experiment.

I. Acclimation of enzyme activity to salinity

The average level of leucine aminopeptidase activity changes in response to differing salinities. The activity of individuals, irrespective of *Lap* phenotype, is high in animals acclimated to normal oceanic salinity (33 o/oo) and low in animals acclimated to low salinity (15 o/oo). Although there is some difference in the course of the acclimation of enzyme activity between the experiments performed in 1976 and 1977 (Figure 4a and b), this is thought to be due to differences in ambient temperatures between the two experiments. In both cases, there is a diminution of *Lap* activity in the first few days of exposure to lowered salinity, followed by a rise in activity to high control levels, and then a steady diminution of activity to a lower steady-state level. Also in both cases, the ultimate decline to lower enzyme activity occurred between nine and ten days. Since microdensitometry of tissue sections during the course of these experiments is still in progress, it is not yet known whether different cell types have different time courses of acclimation which might produce the complex acclimation patterns observed in assayed enzyme activity. Where this has been done (below), total enzyme activity only generally reflects the individually different time courses of the change of *Lap* activity in the different cell types.

This lower leucine aminopeptidase activity in low salinity is apparent in the digestive tubule sections illustrated in Figures 3A (control) and 3B (15 o/oo acclimated). From a visual inspection of tissue sections during the course of acclimation to low salinity, no substantial cell damage was apparent.

The response in enzyme activity to exposure to high salinity is nearly instantaneous (Fig. 5). A sharp rise in enzyme activity occurs over the first few hours of exposure to high salinity, and the rise in enzyme activity is correlated with the rise in blood osmolality (Koehn and Bayne, unpublished). Although a high equilibrium level of enzyme activity is attained within 24 hours of exposure to high salinity (Fig. 5), there are substantial cellular changes in the amount and distribution of *Lap* activity during the ensuing seven to eight days of exposure. These changes appear as a shallow valley in enzyme activity between day 2 and day 7 (Fig. 5), though these values are not significantly different from higher

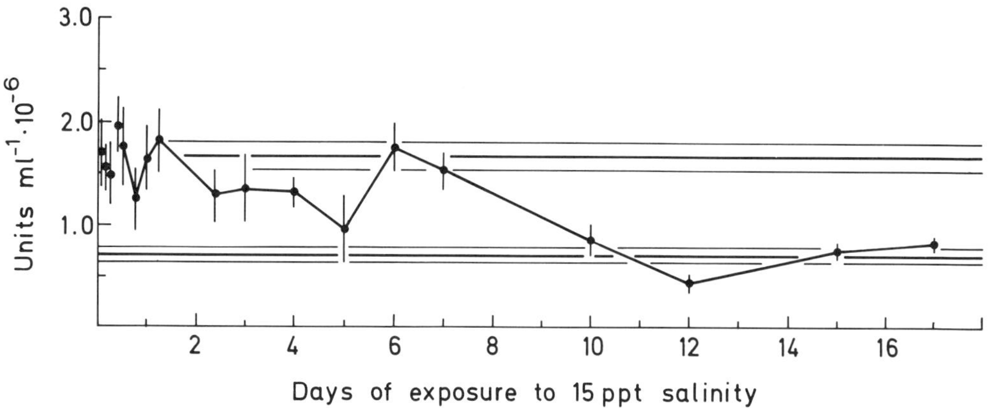

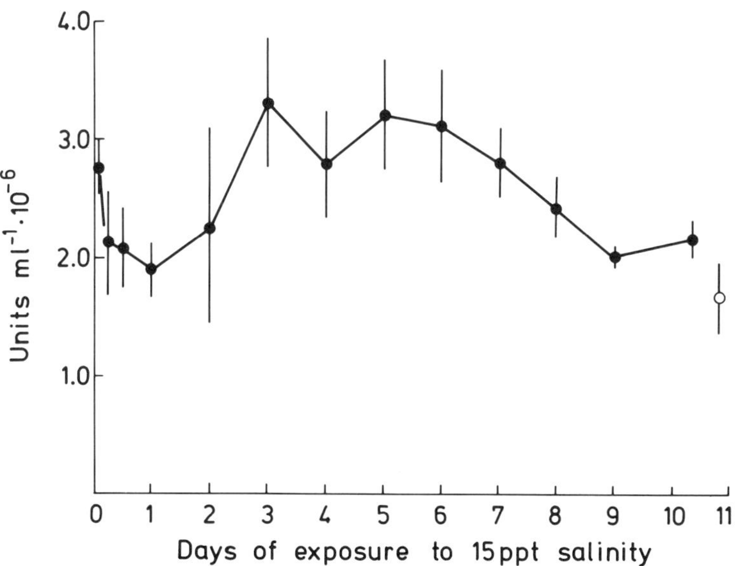

Fig. 4. The change in assayed leucine aminopeptidase activity during acclimation from 33 to 15 ppt salinity. Parallel lines represent average mean ± s.e. of controls (33 ppt, above and 15 ppt, below) during experiment. Experiment in 1976. B. Same as A except single 15 ppt control assay shown at 11 days and experiment done in 1977.

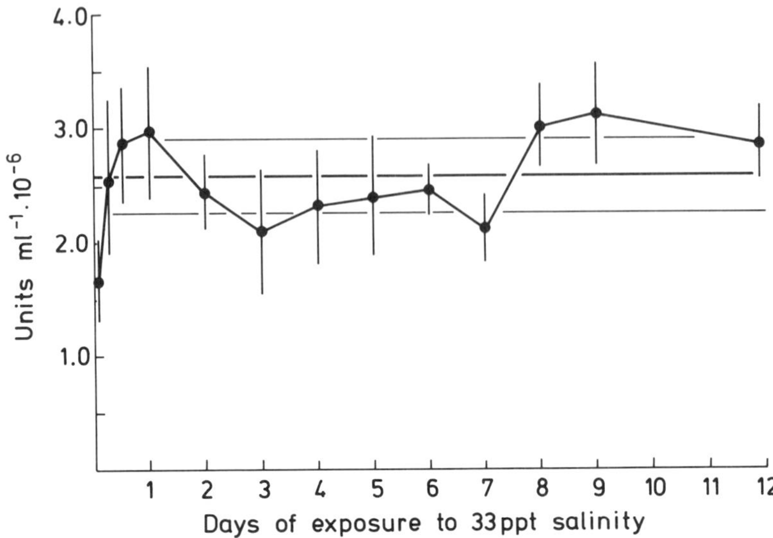

Fig. 5. The change in assayed leucine aminopeptidase activity during acclimation from 15 to 33 ppt salinity. Parallel lines represent average mean ± s.e. of 33 ppt control determined during course of experiment.

levels of enzyme activity that are observed both before and after this period.

Total assayed enzyme is a composite of underlying differences in the acclimation pattern of leucine aminopeptidase activity in individual cell types (compare Figs. 5 and 6). In both digestive tubule cells and the cytoplasm of gut epithelium, there is a very rapid increase in enzyme activity during the first 24 hours of exposure to high salinity, but no change in activity localized to the brush border of gut epithelial cells. There is in fact a substantial "overshoot" of activity in digestive tubule and gut epithelial cells (Fig. 6). It is not presently known whether this instantaneous increase of enzyme is due to enzyme activation or *de novo* synthesis. The lack of a response in the brush border would suggest that the function of *Lap* at this site is different from other sites and may only reflect a lower turnover rate as the enzyme is passing from its site of synthesis in the gut epithelial cells to the digestive lumen.

During the course of acclimation to high salinity there is substantial destruction of the gut epithelium (Fig. 3F). This is accompanied by marked decline of enzyme activity in the brush border between one and seven days (Fig. 6). As the brush border reforms, there is a simultaneous increase in all tissue types to higher equilibrium levels (Fig. 6) after about 9 days.

It is apparent from the above experimental results that levels of *Lap* activity can be affected by changes in environmental salinity. When there is a decrease in salinity, there is a gradual diminution of enzyme activity, but when salinity is increased, the increase in enzyme activity is nearly instantaneous.

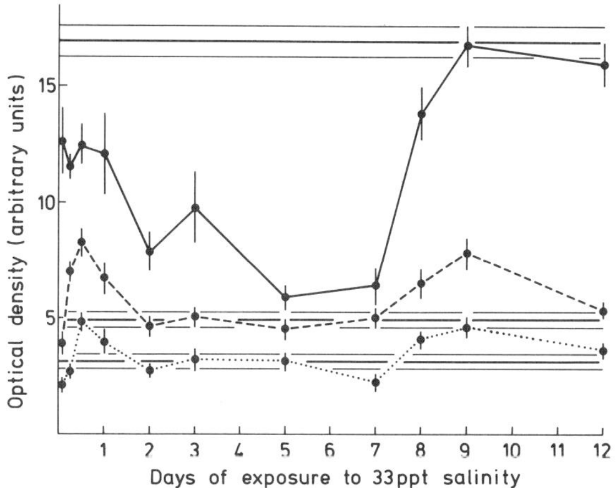

Fig. 6. The change in leucine aminopeptidase activity in digestive tubule cells (---), cytoplasm of intestinal epithelial cells (....) and brush border of intestinal epithelial cells (——) during acclimation from 15 to 33 ppt salinity. Enzyme activity was determined by microdensitometry at 500 nm and is expressed in arbitrary absorption units. Parallel lines are 33 ppt controls as given in Figure 5.

The rate of acclimation to a rise in environmental salinity will be a function of the rate of increase in the cellular free amino acid pool. The ability of an organism to become isoosmotic with environmental salinity will depend on the speed with which the population of free amino acids can be increased. The faster the rate of increase, the less osmotic damage will occur as a consequence of the change in salinity. The change in the free amino acid pool then must depend upon the rate at which it is provisioned, and it is at this point that we suppose that the leucine aminopeptidase enzyme is important. Whatever the mechanism of increase in enzyme activity, we can conclude that more free amino acids are being produced by the *Lap* enzyme at high environmental salinities than at low environmental salinities. The nearly instantaneous rate of increase of enzyme activity would be required by any component within the coordinated physiological response which must occur to increase the free amino acid pool. The *Lap* enzyme is no doubt only one of a series of enzymes which contribute to this response. The time course of increase in *Lap* activity follows that of the acclimation of the free amino acid pool, which occurs over about 24 hours. At high salinities, the concentration of cellular free amino acids will be about double and differ from the extracellular concentration by orders of magnitude. Therefore, we might also expect that different *equilibrium* levels of enzyme activity will be required to maintain the free amino acid pool at different concentrations.

II. Enzyme activity in natural populations

Having demonstrated that *Lap* enzyme activity responds to environmental salinity, it is important to examine whether or not differences occur in natural populations that are exposed to different salinities and whether or not the pattern of difference in enzyme activity relates to the pattern of differences in genetic composition. This is, after all, our ultimate concern. On Long Island there are large differences in average enzyme activity between Atlantic Ocean and Long Island Sound populations. The Shinnecock and Stony Brook study sites are representative of these extremes (Table 1). This difference in average activity (irrespective of phenotype) has been observed, without exception, during the sampling of these populations over the past two and one half years. This difference is maintained despite substantial seasonal variations in average enzyme activity (Koehn, 1978). These may occur as a consequence of the seasonal variations in the degree to which protein serves as an energy source in this species. The difference in average enzyme activity is not merely due to the presence in each population of differing frequencies of phenotypes of significantly different enzyme activity, since significant differences between the two populations can be observed in the enzyme activity for individual phenotypes (Table 1).

Not only do the two study populations, each exposed to differing salinities, differ in the average level of enzyme activity, but the transition from high to low enzyme activity corresponds spatially to the transition in the frequency of *Lap*^{94} from high to low at the entrance to Long Island Sound (Fig. 7). The form of the allele fre-

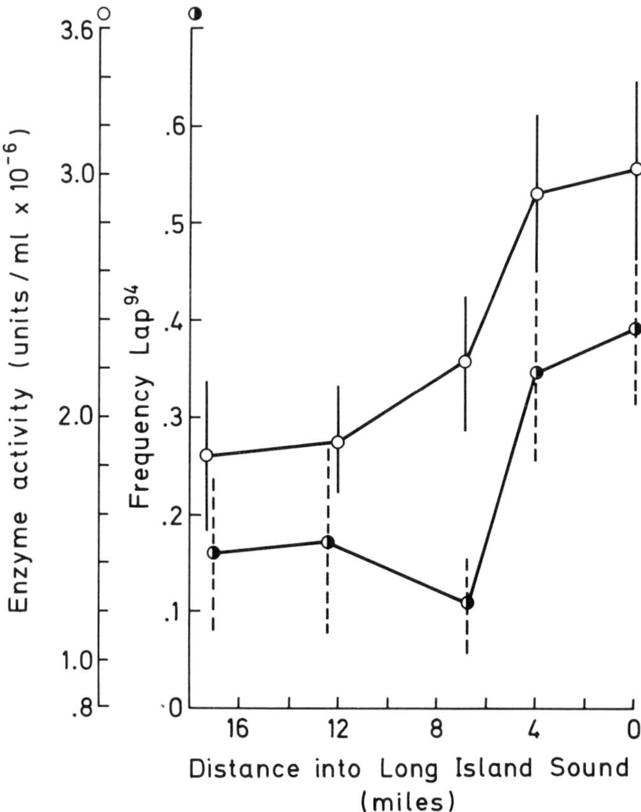

Fig. 7. Clinal variation in the frequency of the *Lap*^{94} allele (0) and assayed enzyme activity (0) among populations at the eastern entrance to Long Island Sound on the Long Island Shore (see Fig. 1). All points are the mean ± s.e.

quency cline in Figure 7 is somewhat different than that illustrated in Figure 2, since these samples were taken in a different season. Average sample sizes were also much smaller due to the infeasibility of assaying large samples.

The difference in enzyme activity between the two Long Island populations occurs in the digestive tubule cells and cytoplasm of the gut epithelium. Surprisingly, it does not occur in the brush border of the epithelial cells (Table 4). The absence of a difference in the brush border in these natural populations must be verified in view of the differences in brush border activity observed in acclimation experiments (Fig. 6).

Table 4. Leucine aminopeptidase activities in different cellular localizations in two populations of *Mytilus edulis*. Enzyme activity is expressed in arbitrary absorption units, determined by cellular microdensitometry.

	Stony Brook	Shinnecock	t Statistic
Digestive cells	2.108	6.463	20.20[a]
Gut epithelium	2.403	3.844	6.24[a]
Gut brush border	17.165	17.832	1.53

[a] $P < .001$

F. Is phenotypic diversity at the Lap locus adaptive?

Spatial variation of allele frequencies at the *Lap* locus has been attributed to the role of natural selection (Koehn et al., 1976; Milkman and Koehn, 1977). Starting from the premise that variation at a locus is of population genetic interest because it is consistent with our intuitive judgment of how natural selection may produce such patterns, we have been forced to take lengthy excursions into biochemistry and physiology in order to attempt to demonstrate how selection could in fact be occurring. Since neither the biochemical function nor the physiological role of the leucine aminopeptidase enzyme was known, estimates of these had to be established, so that the underlying selective mechanism could ultimately be demonstrated. Though we have gone a long way toward fulfilling this goal, it is still not possible to demonstrate unequivocally the selective maintenance of the Lap polymorphism. Nor is it possible at this point to understand the bases for the differentiation of allele frequencies among populations exposed to differing salinities. There are, however, certain conclusions which can be made which bear on the ultimate answer to this question. Phenotypic diversity exists among *Lap* phenotypes in enzyme activity in natural populations and in the degree to which individual phenotypes are sensitive to the presence of inhibiting cations. There are also substantial differences in enzyme activity among populations exposed to different environmental salinities. Enzyme activity is high in high salinities and low in lower salinities. Salinity-correlated activity differences can be acclimated in response to experimental variations in salinity. The change in enzyme activity is a regulated response to salinity variations, presumably due to the role of the *Lap* enzyme in providing free amino acids for osmoregulation. Thus, there is an intercorrelation among salinity, enzyme activity, and allele frequency in populations distributed over a gradient of environmental salinity. This observation would suggest that environmental salinity not only stimulates a response in enzyme activity within generations, but may serve as a source of natural selection in the intergeneration establishment of allele frequency differences among populations. Finally, there is a high frequency (approximately 0.55) of the allele with the highest enzyme activity (Lap^{94}) in populations exposed to the highest salinity and having the highest average enzyme activity. The converse is also true. Thus, there is both a phenotypic (activity) and genotypic (allele frequency) response to salinity variations.

These observations collectively constitute a coherent picture of a seemingly adaptive polymorphism, wherein high enzyme activity is favored in high salinity environments. Nevertheless, many important elements are lacking in a total explanation for the *Lap* polymorphism. For example, selection in favor of high activity *per se* cannot alone explain the maintenance of the polymorphism. Some properties of the Lap^{98} and Lap^{96} alleles must insure their higher frequency in low salinity environments. These are presently unknown. There is some evidence that spatial variation of allele frequencies is influenced by environmental temperatures (Koehn et al., 1976); and when enzymological characterization of the individual alleles has been completed, these properties will presumably emerge.

If the high average enzyme activity and the high frequency of a high activity allele in normal oceanic salinity reflects the need for a more rapid rate of free amino acid production for osmoregulation, then the following expectations can be stated. First, since the free amino acid pool has certain dominant residues (e.g. glycine, taurine), acclimation to elevated salinity and rise in *Lap* activity should be accompanied by a rise in the activity of various transaminase enzymes. Second, the acclimation rate of the free amino acid pool and *Lap* activity should be faster in individuals with the Lap^{94} allele.

The location of the leucine aminopeptidase enzyme in distinctly different cell types would suggest that it has more than one physiological function and that the interaction of selection and phenotypic diversity may in reality be a consequence of synergistic physiological events. That is, while Lap^{94} might be more highly favored in higher environmental salinities because it assists in a high rate of salinity acclimation, other *Lap* alleles may be favored in other ecological circumstances because of their unique properties in some other physiological function such as digestion. It is probable that the mechanism of maintenance of this polymorphism is more complex than we have either described or speculated.

Although there is some evidence of viability differences among the *Lap* phenotypes (Milkman and Koehn, 1977), it has not yet been established that the described phenotypic diversity in enzyme function exists as variations in Darwinian fitness. Phenotypic diversity in molecular function is not equivalent to fitness differences among phenotypes. Moreover, very dramatic functional differences do not necessarily imply large fitness differences. There is no obvious relationship between a phenotype scale (e.g. enzyme activity) and a fitness scale.

G. Concluding Remarks

The leucine aminopeptidase locus is illustrative of the different levels of response an environmental challenge may elicit in organisms. The initial response of the species to a change in salinity is valve closure, thus securing a physiological "period of grace" (Pierce, 1971). When this is not an effective adaptation, a variety of physiological responses occur. One of these is a change in levels of leucine aminopeptidase activity. When the required adaptation falls outside the range of the physiological response, an apparent genetic consequence is manifested as the spatial heterogeneity of allele frequencies correlated with spatial variations in salinity. It is interesting why a system such as the *Lap* locus, well regulated in a coordinated physiological response, exhibits allelic variants that are apparently an adaptation to the same environmental challenge served by the regulation. This might imply that the rate of acclimation to osmotic stress would be different among the individual *Lap* phenotypes. In such a case, genic variations at a structural locus would allow the regulation rate to be maximized. When increased Lap enzyme activity is required, regulation would increase the effective enzyme activity by both *de novo* synthesis and/or enzyme activation. An allelic form of the enzyme with a higher hydrolytic rate would permit the regulation rate to proceed even faster.

Although similar phenotypic diversity may exist at homologous *Lap* loci in other bivalve species and extrapolation of our results in *Mytilus edulis* to these species seems justified in view of the similar interspecific variation of allele frequencies (Koehn and Mitton, 1972), it is not likely that these results should be taken to characterize all, or even a majority, of polymorphic enzyme loci. Leucine aminopeptidase is involved in basic physiological processes whose responses are mediated by changes in the external environment. These responses are immediate. Only a very small number of enzyme loci that are typically examined to compute "average heterozygosity," are analogous to *Lap*. These include some of the variable substrate enzymes, but probably not enzymes involved in intermediary metabolism. Until more detailed biochemical and physiological investigations of polymorphic enzymes are performed, it seems advisable to view phenotypic diversity of *Lap* in *M. edulis* as an exception, rather than the rule.

In view of the great amount of effort we have expended in order to elucidate certain functional properties of the *Lap* enzymes, we should consider whether or not this is an appropriate strategy for ultimately understanding the maintenance of enzyme polymorphism in nature. However detailed the characterization of genetic variation of an enzyme, and however compelling the evidence may be for selective maintenance of the variation, it does not necessarily follow that selection occurs only at this single locus. Even the intercorrelations of salinity, enzyme function and allele frequency variations would not justify such a conclusion. Any assignment of fitness values to phenotypes must be somewhat artificial since all enzymes function in a complex metabolic system. The response of the free amino acid pool in *M. edulis* is a coordinated physiological response, which must additionally involve other peptidases, proteinases, and transaminases. Indeed, we have previously presented evidence that there is an epistatic interaction between leucine aminopeptidase and another aminopeptidase (Mitton and Koehn, 1973). Diversity of molecular function would be expected with both single locus selection and the epistatic interaction of individual phenotypes at a series of loci.

By singling out a single locus from a complex network of biochemical reactions, we are attempting to identify certain key elements of this network which will be more highly correlated with ecological variables than other elements of the network.

For years, ecological physiologists have been exploring the integrated response of biochemical reactions to environmental variation without consideration of the enzymological diversity in the form of polymorphism that might underlie the pattern of physiological response. Ecological geneticists, on the other hand, have investigated enzyme polymorphism at loci of largely unrelated function, with the hopes of discovering their adaptive significance from single and joint properties of allele frequency variations. While the "single locus approach" provides important information on the nature of functional differences of allozymes, these differences alone cannot explain the maintenance of enzyme polymorphism outside of a physiological context. The coordination of physiological and genetic studies seems to be a mutually beneficial undertaking and a major challenge for the future.

Acknowledgements. This study has involved a great many people, each of whom has contributed significantly in one way or another to the results reported here. The direction of this study has relied heavily on descriptive population genetics published earlier with Drs. Roger Milkman, Jeffrey B. Mitton, and Mr. Francis Turano. I am especially grateful to Mr. Hans H. Lassen and F. Turano for use of the data illustrated in Figure 2. Dr. J. Peter W. Young performed many of the biochemical investigations on this enzyme, and we have been assisted by the guidance of Dr. Norman Arnheim and the laboratory help of Ms. Doris Wiener. Drs. Young and Arnheim prepared antibodies to purified Lap.

I am especially grateful to the warm hospitality of the Institute for Marine Environmental Research, Plymouth, England, where all of the acclimation and cytochemical studies of Lap were done in collaboration with Drs. Brian L. Bayne and Michael Moore. The great number of enzyme assays that had to be done to characterize natural populations were assisted by Mr. Frederick Immerman, David Innes, Allen Lamb and Patrick Gaffney.

This research was supported by U.S.P.H.S. Career Award GM-28963, U.S.P.H.S. Research Grant GM-21133, NSF Research Grant BMS 74-02522, and a NATO Senior Fellowship in Science. This is contribution number 240 from the Program in Ecology and Evolution at the State University of New York at Stony Brook.

References

Adibi, S.A., Mercer, D.W.: Protein digestion in human intestine as reflected in luminal, mucosal, and plasma amino acid concentrations after meals. J. Clin. Invest. 52:1594-1596. (1973).

Ayala, F.J. (ed.): Molecular Evolution. Sunderland, Mass.: Sinnauer Assoc., Inc., 1976, p. 277.

Barrett, A.J., Dingle, J.T. (eds.): Tissue Proteinases. Amsterdam: North-Holland Publ. Co., 1971, p. 353.

Bayne, B.L., Widdows, J., Thompson, R.J.: Physiology II. In Marine Mussels: Their Ecology and Physiology. (ed. B. Bayne). Cambridge: Cambridge University Press, 1976, pp. 207-260.

Bishop, S.H.: Nitrogen metabolism and excretion: Regulation of intracellular amino acid concentrations. In Estuarine Processes, Vol. I. (ed. M. Wiley). New York: Academic Press, 1976.

Bricteux-Gregoire, S., Duchateau-Bosson, G., Jeuniaux, C., Florkin, M.: Constituants osmotiquement actifs des muscles adducteurs de *Mytilus edulis* adaptee a l'eau ou a l'eau saumatre. Arch. Int. Physiol. Biochem. 72:116-123. (1964).

Bryant, E.H.: On the adaptive significance of enzyme polymorphism in relation to environmental variability. Amer. Natur. 108:1-19. (1974).

Day, T.H., Hillier, P.C., Clarke, B.: Properties of genetically polymorphic isozymes of alcohol dehydrogenase in *Drosophila melanogaster*. Biochem. Genet. 11:141-153. (1974a).

Day, T.H., Hillier, P.C., Clarke, B.: The relative quantities and catalytic activities of enzymes produced by alleles at the alcohol dehydrogenase locus in *Drosophila melanogaster*. Biochem. Genet. 11:155-165. (1974b).

Day, T.H., Needham, L.: Properties of alcohol dehydrogenase isozymes in a strain of *Drosophila melanogaster* homozygous for the *Adh*-slow allele. Biochem. Genet. 11:167-175. (1974).

Ford, E.B.: Genetic Polymorphism. All Soul's Studies. London: Faber and Faber, 1965, p. 101.

Gibson, J.: Enzyme flexibility in *Drosophila melanogaster*. Nature 227:959-960. (1970).

Hedrick, P.W., Ginevan, M.R., Swing, E.P.: Genetic polymorphism in heterogeneous environments. Ann Rev. Ecol. Syst. 7:1-32. (1976).

Kim, M.S., Brohpy, E.J.: Rat intestinal brush border membrane peptidases. I. Solubilization, purification, and physicochemical properties of two different forms of the enzyme. J. Biol. Chem. 251:3199-3205. (1976).

Kim, Y.S. McCarthy, D.M., Lane, W., Fong, W.: Alterations in the levels of peptide hydrolases and other enzymes in brush-border and soluble fractions of rat small intestinal mucosa during starvation and refeeding. Biochem. Biophys. Acta 321:262-273. (1973).

Kim, Y.S., Kim, Y.W., Sleisenger, M.H.: Studies on the properties of peptide hydrolases in the brush-border and soluble fractions of small intestinal mucosa of rat and man. Biochem. Biophys. Acta 370:283-296. (1974a).

Kim, Y.S., Nicholson, J.A., Curtis, K.J.: Intestinal peptide hydrolases: Peptide and amino acid absorption. Med. Slin. N. Amer. 58:1397-1412. (1974b).

Koehn, R.K.: Esterase heterogeneity: Dynamics of a polymorphism. Science 163:943-944. (1969).

Koehn, R.K.: Biochemical aspects of genetic variation at the LAP locus in *Mytilus edulis*. In The Genetics of Marine Organisms. (eds. B. Battaghlia and J.A. Beardmore). New York: Plenum Press, 1978.

Koehn, R.K., Mitton, J.B.: Population genetics of marine pelecypods. I. Ecological heterogeneity and evolutionary strategy at an enzyme locus. Amer. Natur. 106:47-56. (1972).

Koehn, R.K., Milkman, R., Mitton, J.B.: Population genetics of marine pelecypods. IV. Selection, migration and genetic differentiation in the Blue Mussel, *Mytilus edulis*. Evolution 30:2-32. (1976).

Levene, H.: Genetic equilibrium when more than one niche is available. Amer. Natur. 87:331-333. (1953).

Levins, R.: Evolution in Changing Environments. Princeton, New Jersey: Princeton University Press, 1968, p. 120.

Levinton, J.S., Koehn, R.K.: Population genetics of mussels. In Marine Mussels: Their Ecology and Physiology. (ed. B.L. Bayne).

Cambridge: Cambridge University Press, 1976, pp. 357-384.
Lewontin, R.C.: The Genetic Basis of Evolutionary Change. New York: Columbia University Press, 1974, p. 346.
Maroux, S., Louvard, D., Baratti, J.: The aminopeptidase from hog intestinal brush border. Biochim. Biophys. Acta 321:282-295. (1973).
Merritt, R.B.: Geographic distribution and enzymatic properties of lactate dehydrogenase allozymes in the Fathead Minnow, *Pimaphales promelas*. Amer. Natur. 196:173-184. (1972).
Milkman, R., Koehn, R.K.: Temporal variation in the relationship between size, numbers and an allele-frequency in a population of *Mytilus edulis*. Evolution 31:103-115. (1977).
Miller, S., Pearcy, R.W., Berger, E.: Polymorphism at the α-glycerophosphate dehydrogenase locus in *Drosophila melanogaster*. I. Properties of adult allozymes. Biochem. Genet. 13:175-188. (1975).
Mitton, J.B., Koehn, R.K.: Population genetics of marine pelecypods. III. Epistasis between functionally related isoenzymes of *Mytilus edulis*. Genetics 73:487-496. (1973).
Nei, M.: Molecular Population Genetics and Evolution. Amsterdam: North-Holland Publ. Co., 1975, p. 288.
Pierce, S.K.: Volume regulation and valve movements by marine mussels. Comp. Biochem. Physiol. 39A:103-117. (1971).
Piggott, C.O., Fottrell, P.F.: Purification and characterization from guinea-pig intestinal mucosa of two peptide hydrolases which preferentially hydrolyse dipeptides. Biochim. Biophys. Acta 391:403-309. (1975).
Smyth, D.H.: Peptide transport by mammalian gut. In Peptide Transport in Bacteria and Mammalian Gut. Ciba Foundation Symposium. Amsterdam: Elsevier, Excerpta Medica, North-Holland, 1972, pp. 59-70.
Somero, G.N., Low, P.S.: Eurytolerant proteins: Mechanisms for extending the environmental tolerance range of enzyme-ligand interactions. Amer. Natur. 111:527-538. (1977).
Thompson, R.J., Ratcliff, N.A., Bayne, B.L.: Effects of starvation on structure and function in the digestive gland of the mussel (*Mytilus edulis* L.). J. Mar. Biol. Assn. U.K. 54:699-712. (1974).
Thorig, G.E.W., Schoone, A.A., Scharloo, W.: Variation between electrophoretically identifiable alleles at the alcohol dehydrogenase locus in *Drosophila melanogaster*. Biochem. Genet. 13:721-731. (1975).
Vigue, C.L., Johnson, F.M.: Isozyme variability in species of the genus *Drosophila*. VI. Frequency-property-environment relationships of allelic alcohol dehydrogenases in *D. melanogaster*. Biochem. Genet. 9:213-228. (1973).
Wojnarowska, F., Gray, G.M.: Intestinal surface peptide hydrolases: Identification and characterization of three enzymes from rat brush border. Biochim. Biophys. Acta 403:147-160. (1975).
Young, J.P.W., Koehn, R.K., Arnheim, N.: An aminopeptidase from the mussel, *Mytilus edulis*: Purification and properties. (MS in preparation, 1978).

Genes, Enzymes and Hypoxia

W.J. SCHULL, R.E. FERRELL, and F. ROTHHAMMER

A. Introduction

Twenty thousand years or so ago, man appeared in the New World. We do not know how many prehistoric migrants reached the Americas through Beringia on this first occasion, - presumably only a few - nor how many subsequent migrations took place, nor the physical appearance of these individuals (see Laughlin, 1977). It is assumed they were mongoloid and possibly similar to the peoples presently indigenous to northeastern Siberia. Surprisingly quickly after their arrival, descendants of these early migrants had peopled the plains and woodlands of Canada and the United States, filtered through Mexico and the Isthmus of Panama, and had penetrated the southern continent. We do not know to which of the numerous regions of the Americas through which these early nomads passed in this prehistoric hegira they were best suited, but few regions would have been more hostile than the Andean altiplano which they reached twelve thousand years ago (MacNeish et al., 1970; but see also MacNeish, 1971). Ambient temperature can and frequently does vary more than 50°C in the span of 24 hours; little shelter exists from the chilling, parching winds, and the partial oxygen pressure is but 60% of that which obtains at sea level. Normal thermoregulation is seriously challenged; so is the transport and tissue utilization of oxygen. However, man has obviously thrived in the Andes; more than ten million individuals now live there at altitudes of 3000 meters or more (Baker, 1977). This increase in numbers undoubtedly reflects an ability to influence environmental circumstances through the domestication of a wide variety of frost-resistant plants. The discovery of means to store surpluses in the form of herds of alpaca and llama are examples. But has man himself adapted? Has his genetic makeup been systematically altered as a consequence of the rigors of this environment? If so, how is one to demonstrate this fact? Surely much of the genetic contribution to the adaptation of the contemporary Andeans (e.g. the Aymara and Quechua) must have occurred generations, possibly millenia, ago, and thus is inaccessible to study.

B. Transport and Tissue Utilization of Oxygen

Functional adaptation to the hypoxia of altitude, as others have noted (see, e.g., Frisancho, 1975), could occur through modifications in pulmonary ventilation, lung volume and pulmonary diffusing capacity, transportation of oxygen in the blood, diffusion of oxygen from blood to tissue, utilization of oxygen at the tissue level, or some combination thereof. Proof of an adaptive change in any one entails first the demonstration that the variables associated with a particular pathway change systematically with changes in the selective force, in this case oxygen tension. Second, the trait which responds must be one known to be under genetic control. We shall restrict our attention here to changes of a biochemical nature which could affect the transportation and tissue utilization of oxygen.

I. Oxygen transport: general mechanisms. It is common knowledge that hemoglobin is the vehicle by which oxygen is transported from the capillaries of the lungs to the cells of the body; in turn, CO_2 is carried from the tissues to the capillaries of the lung. The intricate details of these transport functions are less well known. A hemoglobin molecule is comprised of four subunits, each of which contains a heme moiety attached to a polypeptide chain. Oxygen can be reversibly bound to an atom of ferrous iron found in each of the four heme moieties; thus, one molecule of hemoglobin can bind four molecules of oxygen. Combination with oxygen of the first heme in a hemoglobin molecule increases the affinity of the second heme for oxygen; oxygenation of the latter increases the affinity of the third, etc. This shifting affinity, presumably secondary to the movement of the beta chains closer together, gives rise to the sigmoidal shape of the oxygen-hemoglobin dissociation curve. The temperature and pH of the suspending medium, serum, also affect hemoglobin's capacity to acquire and release oxygen. If temperature rises or the pH of blood falls a higher partial oxygen pressure is required for hemoglobin to bind a given amount of oxygen. While the former effect is presumably a thermodynamic one, the pH or Bohr effect reflects the fact that deoxyhemoglobin binds hydrogen ions more avidly than does oxyhemoglobin. This buffers venous blood and facilitates the formation of bicarbonate, one of the mechanisms by which tissues dissipate carbon dioxide. It is also known that an intermediate product of anaerobic glycolysis, 2,3-diphosphoglycerate, influences hemoglobin's affinity for oxygen. 2,3-DPG is a highly charged anion which binds to the beta chains of deoxygenated hemoglobin but not to those of oxyhemoglobin; it stabilizes the deoxy conformer and thus lowers the affinity of hemoglobin for oxygen. The residues responsible for binding 2,3-DPG are the two alpha amino groups of the beta chains; hence, 2,3-DPG competes with CO_2 for binding sites.

Energy is required to acquire or release oxygen. Most, if not all of this energy arises through the process of phosphorylation leading to the formation of so-called high energy phosphate compounds; the most important of these is adenosine triphosphate (ATP). The latter is derived from adenosine diphosphate (ADP) either oxidatively within the mitochondria or at the substrate level through glycolysis. The former pathway is aerobic; the latter is not. Both processes, however, depend upon the removal of hydrogen ions from metabolic intermediates. In the aerobic pathway, energy derived from the oxidation of hydrogen ions to water is used to form ATP. This pathway, which is the more efficient producer of ATP, is directly compromised by hypoxia. In the anaerobic pathway phosphorylation depends on the removal of hydrogen ions by the formation of lactate from the pyruvate produced by the metabolism of glucose and glycogen. Glucose can be converted to pyruvic acid either through the Embden-Meyerhof pathway or through the pentose monophosphate shunt.

The human erythrocyte lacks mitochondria, and thus represents an ideal system in which to study glycolysis, uncontaminated by any other interfering pathway. Since virtually all oxygen which reaches the tissues must do so through the red blood cells, the importance of their biochemical activity is self-evident. Normally about 90% of the glucose metabolized in the red blood cell passes through the Embden-Meyerhof pathway; the remainder traverses the pentose monophosphate shunt. The latter is also termed the oxidative pentose path, which might suggest that it too would be compromised by hypoxia. This is not the case. Even though the process is oxidative in the sense that carbon dioxide is produced, oxygen does not enter directly as an electron acceptor. Two enzymes, glucose-6-phosphate dehydrogenase and 6-phosphogluconate dehydrogenase, control the gate to the pentose monophosphate shunt, a matter to which we shall return. It warrants note here that the red blood cell has no internal compartments, and the data on enzyme kinetics and metabolite concentrations are superior to those of most other cells.

II. Tissue utilization of oxygen. Our remarks thus far have been addressed primarily to the transport of oxygen. Tissues differ in their capacity to extract oxygen from the circulating red blood cells and in their apparent sensitivity to oxygen deprivation. The central nervous system, for example, is unusually sensitive to a fall in partial oxygen pressure. The myocardium is greatly dependent upon oxygen, but it is an efficient extractor. The oxygen saturation in the coronary sinus (the venous side of the coronary circulation) at sea level is only 25%; whereas, in the mixed venous blood oxygen saturation is about 70%. Harris and his colleagues (Harris, 1977; Harris et al., 1970; Barrie and Harris, 1976) have examined the effect of chronic hypoxia on the activities of myocardial enzymes in guinea pigs. They find that chronic hypoxia results in a right ventricular hypertrophy in these animals, as it also does in man, along with an increase in the activity of the enzymes of the glycolytic pathway. More specifically, hypertrophy prompts an increase in the activities of glycogen phosphorylase, phosphoglucomutase, hexokinase, glucose phosphate isomerase, glyceraldehyde-3-phosphate dehydrogenase, and phosphoglycerate kinase. All of these enzymes favor reactions that flow through the Embden-Meyerhof pathway. Whether these latter changes also occur in man at high altitude is not known. However, there are some enigmatic observations by cardiac physiologists which are pertinent. Hurtado (1971) has asserted, "We have found that the native subject, in his high environment, has a better average tolerance time to a maximal period of exercise; and somewhat paradoxically his oxygen debt and lactate and pyruvate production are definitely lower, indicating a more aerobic than anaerobic source of energy." This seems at variance with the increase in the activity of enzymes associated with anaerobic glycolysis.

To summarize briefly, the transport and tissue utilization of oxygen, particularly the former, could be challenged by any biochemical change which affects the composition of hemoglobin, the temperature or pH of blood, or the energy sources needed to effect the acquisition and release of oxygen or carbon dioxide either in the red blood cell or tissues. Thus, an alteration in an enzyme such as carbonic anhydrase, which expedites the conversion of carbon dioxide and water to carbonic acid (or the reverse) could affect blood pH; or alterations in the enzymes involved in glycolysis could affect energy resources. Clearly the physiologic and reproductive consequences of these alterations are not readily predicted; the impact of a particular change would presumably depend upon the role of the enzyme involved, the extent of impairment (or enhancement) of normal function, and the availability of alternative metabolic pathways.

C. Multinational Andean Genetic and Health Program

The objective of our study is to appraise the contribution of genetic variability to the anatomical, biochemical and physiological responses of the Aymara, the indigenes of western Boliva, northern Chile, and southern Peru, to the hypoxia of altitude (Schull and Rothhammer, 1977a,b; see Tschopik, 1963, for a cultural account of the Aymara). While this assessment has many aspects, we are concerned here only with those which relate to biochemical variability. Additional observations connected with this study include (1) a complete medical history and general physical examination, (2) oral and dental examinations, (3) anthropometric measurements, (4) pulmonary function tests, (5) resting electrocardiogram, (6) ophthalmoscopic examination including visual acuity, color vision tests and tensiometry, (7) detailed appraisal of cardiovascular status, (8) simple performance tests such as tapping, (9) nutritional, reproductive and residential histories, and finally (10) either an ACD-preserved specimen of venous blood, or a prechloric or trichloracetic acid precipitated specimen, or both.

In the years 1973 through 1975 a multinational group of investigators from Bolivia, Chile, Ecuador, Peru and the United States examined 2525 individuals (Aymara, Mestizos and non-Aymara) who were residents of 14 villages or locales in western Bolivia and northern Chile. Elevations ranged from sea level to somewhat more than 4200 meters. Blood specimens were obtained from about 1700 of these persons. Preliminary processing of these specimens occurred in the field. Once or twice a day the ACD specimens were centrifuged; the plasma was pipetted into one ml cryovials; and the sedimented red cells were washed in saline and centrifuged again. At the end of three washings and sedimentations, the red cells were either suspended in a freezing solution (a mixture of glycerol and sorbitol) and placed in cryovials or packed in the latter without suspending medium. The cryovials containing plasma, packed cells, or cells suspended in freezing solution were then immersed in liquid nitrogen in a Linde LR-30 freezer, and stored until the nitrogen freezer was received in Houston. In the case of the perchloric or trichloracetic acid specimens, venous blood was drawn into 5 ml vacutainers containing 143 units of sodium heparin. Within 30 seconds or so after completion of the blood-letting, 1 ml of the heparin-blood mixture was precipitated in a sealable plastic tube using a 6.7% solution of trichloracetic acid (TCA) or a 0.6 N solution of perchloric acid. The resulting mixture was shaken quickly and immersed in liquid nitrogen within a minute or so of precipitation. These were so stored, along with the ACD-specimens, until transferred to a larger nitrogen freezer upon arrival in Houston.

The biochemical methods used to identify the isozymic variants were conventional, and their description can be found elsewhere (Bertin et al., 1977; Ferrell et al., 1978a). An extended account of our findings is given in Ferrell et al., (1978b,c).

D. Biochemical Variability among the Aymara

Four lines of biochemical evidence appear especially pertinent to the assessment of the role of genetic factors in the adjustment of the Aymara to the hypoxia of altitude. These are the frequency and fate of (1) abnormal variants of human hemoglobin, (2) deficiencies in those enzymes intimately relatable to the transport or tissue utilization of oxygen, (3) rare isozymic variants - presumably recent mutations in most instances - at loci associated with enzymes involved in glycolysis or oxygen transport and utilization, and (4) changes in the frequencies of common alleles at those loci associated with glycolysis and oxygen transport and utilization, presumably indicative of functional dissimilarities. Through the assessment of the frequencies of a number of biochemical and immunological genetic markers, we can identify any non-indigenous contribution to the present gene pool of the peoples of interest to us. Since the villages are generally small, stochastic effects might occasionally be identified erroneously as systematic ones. To avoid this possibility, a second array of genetic markers, not identified with glycolysis nor known to be involved in oxygen transport, was measured. These enzymes are presumably not under systematic, selective pressure associated with changes in partial oxygen tension, but should reflect various stochastic elements such as drift, founder effect, and migration.

I. Frequency and fate of abnormal variants of hemoglobin: We have observed no electrophoretic abnormalities of hemoglobin among 1694 blood specimens, nor are we aware that such abnormalities among high altitude indigenes have been reported by others. However, hemoglobin variants not leading to a net charge change would not be recognized in the absence of associated clinical difficulties; and in human hemoglobin, only about one random amino acid substitution in three would be expected to lead to a charge change. Japanese investigators-who have striven mightily to ascertain the frequency of hemoglobin variants through peptide mapping, a more discriminating process than electrophoresis alone-have reported 19 kindreds with variants among 50,000 blood samples, or about one variant in every 3000 determinations (Yanase et al., 1968).

Clearly, given the limited scope of our information, we can only speculate on the fate of hemoglobin variants at this altitude. Bellingham (1976) has noted that an abnormal hemoglobin could give rise to an abnormal oxygen dissociation curve for any one or combination of the following reasons: it leads to an intrinsic abnormality in the dissociation curve, to an altered interaction of hemoglobin with 2,3-DPG, or to an altered Bohr effect. Most of the abnormal hemoglobins known to have an altered oxygen affinity have an increased, rather than decreased, affinity; thus, they give up oxgyen to the tissues less readily than normal hemoglobin. This is a distinct disadvantage in an oxygen deprived environment, particularly for a pregnant woman whose fetus is dependent upon her blood supply for oxygen. Most of these defects which lead to an increased affinity appear to be associated with a compensatory increase in circulating erythrocytes (Nagel and Bookchin, 1974; Bellingham, 1976). This would seem to be an adjustment of dubious value in a population already polycythemic, i.e., one where the number of circulating erythrocytes is substantially larger than the number seen in coastal populations. Presumably, there is a limit to the viscosity which blood can attain and not inordinately increase the prospects of an infarct at an early age. Thus, it is difficult to see how any of the presently recognized abnormalities of hemoglobin, except for a very few which lead to a loss in affinity, could improve oxygen transport at altitude. These latter abnormalities, however, present problems as well - mostly in the acquisition of oxygen in lung capillaries. Cyanosis is a common consequence of this limitation in oxygen binding. We saw no individuals who on clinical grounds appeared to have an abnormal hemoglobin.

II. <u>Frequency and fate of enzyme deficiencies:</u> The glycolytic intermediate 2,3-diphosphoglycerate substantially affects the affinity of hemoglobin for oxygen (Benesch and Benesch, 1969). Deficiencies in the enzyme 2,3-diphosphoglycerate mutase, which catalyzes the reaction producing this intermediate, are known to occur (Schroter, 1965). Similarly, deficiencies have been identified in other enzymes associated with glycolysis such as pyruvate kinase, glucose-6-phosphate dehydrogenase, hexokinase, glucose phosphate isomerase, triosephosphate isomerase, phosphoglycerate kinase, phosphofructokinase, and 6-phosphogluconate dehydrogenase (for a recent review, see Valentine and Tanaka, 1972). Without exception, these deficiencies are associated with compromising hemolytic diseases varying in severity. To our knowledge no such deficiency has ever been reported among high altitude natives, nor did we find any. Pyruvate kinase deficiency has been seen, however, in Peru (F. Yen, personal communication). We encountered only 13 cases of anemia (as judged by hemoglobin concentration and hematocrit) among the 1884 individuals we examined who resided at altitudes above 3000 meters. No one of these persons exhibited a clinical picture which was unusual; all seemed readily explicable in terms of nutritional deficiencies (iron), and in no instance were the electrophoretic findings unexpected.

Again, at these altitudes, the fate of individuals homozygous for a mutant which leads to the deficiency of an enzyme in the Embden-Meyerhof or pentose monophosphate shunt pathways can only be conjectured. None of these speculations leads, however, to the conclusion that a deficiency would be advantageous; indeed, it is hard to believe that most individuals so unluckily born would long survive.

III. <u>Frequency and fate of rare, isozymic variants:</u> Here, finally, are data which are provocative. Table 1 sets forth our findings in terms of rare variants for 20 enzymes or serum proteins, some generally monomorphic in human populations and others polymorphic. But first let us consider the roles currently ascribed to the enzymes involved in glycolysis or oxygen transport. Some of the enzymes or serum proteins listed in Table 1 are not presently thought to be involved in any of these processes. These are acid phosphatase, ceruloplasmin, esterase D, haptoglobin, malate dehydrogenase, peptidase, and transferrin. Among the remaining enzymes, at least five are known to influence glycolysis directly or indirectly through participation in a feedback system. These are glucose-6-phosphate dehydrogenase (G-6-PD), 6-phosphogluconate dehydrogenase (6PGD), phosphohexose isomerase (PHI), lactic dehydrogenase (LDH), and adenylate kinase (AK). Three others, adenosine deaminase, isocitrate dehydrogenase and phosphoglucomutase, may also be involved. As indicated earlier, G-6-PD controls the flow of glucose from the early steps of glycolysis into the pentose monophosphate shunt. This diversion is required for the production of metabolically important NADPH, but at the expense of 2,3-DPG. Under circumstances associated with altitude adaptation (e.g., elevated levels of 2,3-DPG) the flow of metabolites through the pentose monophosphate pathway may be affected by 6-PGD due to competitive inhibition of 6-PGD by 2,3-DPG. Beutler et al., (1974) have shown this inhibition to occur at physiological levels of 6-phosphogluconic acid. PHI appears to exert its influence on glycolysis through its metabolic interaction

Table 1. The frequency of rare phenotypes at some 20 genetic loci in the Aymara of Bolivia and Chile

Enzyme or serum protein	Number of individuals tested	Number of individuals with variant phenotypes	Number of distinct variants
acid phosphatase	1694	0	0
adenosine deaminase	1695	0	0
adenylate kinase	1695	10 (10 2-1)	1
esterase D	1693	3	1
glucose-6-phosphate dehydrogenase	1012	16 (4 AB, 12 V)	2
isocitrate dehydrogenase	1672	1	1
lactate dehydrogenase	1695	11	2
malate dehydrogenase	1694	0	0
peptidase A	1621	1	1
peptidase B	1695	2 (2 2-1)	1
peptidase C	1694	0	0
phosphoglucomutase-1	1693	8	1
phosphoglucomutase-2	1694	0	0
6-phosphogluconate dehydrogenase	1692	25 (18 AB, 7 V)	2
phosphohexose isomerase	1691	3	1
albumin 5.0	1681	4	1
albumin 8.3	1692	4	1
ceruloplasmin	1696	0	0
haptoglobin	1672	0	0
hemoglobin	1694	0	0
transferrin	1702	1	1
TOTALS	34767*	89*	16*

*Note: These totals include the two albumin determinations.

Frequency of "rare" phenotypes 2.56/1000 determinations
Frequency of "novel" phenotypes 2.58/1000 determinations

with hexokinase (HK); the latter is the controlling enzyme for the entry of glucose into glycolysis in normal erythrocytes. The activity of HK is, in part, controlled by the level of glucose-6-phosphate, the product of the HK reaction (Rose and O'Connel, 1964). Arnold, et al., (1973) have shown that 2,3-DPG acts as an inhibitor of PHI in the direction F6P→G6P. Thus, elevated levels of 2,3-DPG may activate the HK reaction by relieving HK inhibition by glucose-6-phosphate. The action of LDH is more indirect; the serum lactate:pyruvate ratio, which is influenced by the action of LDH, in turn influences the NAD:NADH (nicotinamide adenine dinucleotide: dihydronicotinamide adenine dinucleotide) ratio. It can thereby affect the activity of glyceraldehyde-3-phosphate dehydrogenase by limiting the available NAD. Finally, AK functions in the production of ADP from adenosine monophosphate (AMP) generated through the purine salvage pathway. ADP is then further phosphorylated to ATP through the action of glycolysis.

Among the 1700 individuals tested, 89 had rare isozymic variants. No less than 65 of these have electrophoretic forms of AK, G-6-PD, LDH, PHI, or 6-PGD not previously associated with high altitude natives. Eight new, or at least uncommon, isozymic forms were encountered. Judged by electrophoretic mobility, five of these appear to be heretofore unrecognized. What should be made of these observations? Are there more or fewer rare variants than experience elsewhere would lead one to expect? Are these rare isozymes disadvantageous, as might be expected, and will soon be eliminated, or are at least some of them advantageous and portend the future genetic constitution of these villages? More specifically, what can be said about the fate of these variants?

As yet, our data are too limited to say much; however, two types of studies seem relevant. First, it would be useful to characterize the kinetic parameters of these new forms. While the evidence thus gained is indirect, in terms of their fate, it could be compelling. Quantitative differences have been shown to exist between the common phenotypes seen at the G-6-PD, 6PGD and AK loci (Harris, 1976). It is reasonable to presume, therefore, that these new variants will not be functionally equivalent to the phenotype associated with homozygosity for the most common allele. What may happen in the case of LDH or PHI is uncertain; no data seen to exist. Second, appraisal of the physiological fitness of carriers of these mutant forms, contrasted with their "normal" siblings and the "general" population of which they are members could be informative. Of special interest, of course, is oxygen consumption in the presence and absence of the mutant forms of these enzymes. This can be measured under laboratory circumstances and can be inferred under day-to-day activity levels. Oxygen consumption bears a fairly simple relationship to heart rate; indeed, this relationship is linear in most individuals; although the slope may vary from person to person (see Astrand and Rodahl, 1970). While it is difficult to monitor oxygen consumption under "normal" circumstances, it is not difficult to monitor heart rate. If, as we have asserted, oxygen consumption is linearly related to heart rate, and if one can assume oxygen consumption at a given heart rate is the same in the laboratory as elsewhere, it would not be difficult to estimate oxygen consumption on an hourly or daily basis through the laboratory appraisal of oxygen consumption at heart rates encountered in continuous monitoring of cardiac activity.

It can, of course, be asked why must one resort to measures of physiological fitness when a measure of reproductive fitness is desired. Clearly, if these mutant forms are incompatible with survival to the age of reproduction, or seriously compromise reproduction thereafter, physiological fitness has a heuristic interest but perhaps not one of population genetic import. If, however, the mutation is consistent with some level of fertility, the estimation of reproductive fitness is a more formidable chore, fraught with many pitfalls (see Weiss and Schull, 1977). These include the large sampling moments associated with estimates of reproductive performance based upon small sample sizes as well as biases such as the dependence of the probability of selection of a sibship upon the number of individuals with the variant phenotype and the correlation between one's reproductive performance and the size of the family from which he or she stems.

To return to Table 1, note that in this population two to three individuals in every thousand have a rare or novel phenotype at least one of the 20 loci studied. Unusual phenotypes (rare variants) were seen at 13 of 20 loci. Comparison of these values with the results of other studies has numerous pitfalls, however.

The same array of loci is rarely studied. It is generally not clear whether the same screening techniques have been used. Most samples involve some related individuals, and thus the number of independent genes which have been assessed at a locus is not clear. Despite these caveats, several comparisons are possible. In a recent study in Iran Kirk and his colleagues (Kirk, et al. 1977) encountered three to four individuals with a rare phenotype in every thousand. They measured the same array of enzymes set forth in Table 1 save for esterase D. Rare phenotypes were seen at 9 of 19 loci. Neel et al. (1976) have reported the frequency of rare phenotypes per 1000 determinations in Micronesians to be approximately one in a thousand. Again, the array of enzymes measured was different, but if their observations are restricted to the set most comparable to our own (they do not report esterase D, G-6-PD or peptidase C variants), the frequency of rare phenotypes remains essentially unchanged. However, only three of the 17 loci studied contribute to the number of rare phenotypes seen. Tanis et al. (1973) report 3.2 individuals per thousand Yanomama to have a rare variant as defined here. Finally, Harris et al. (1973) find a frequency of 2.9 rare variants among Caucasians living in the British Isles, Western Europe and the United States. Thus, published estimates of rare phenotypes differ by about a factor of three; our estimate is neither the largest nor the smallest. Whether this three-fold variation has biologic significance is moot. Neel et al. (1976) argue that it has. We are less intrigued by the magnitude of the variation than by the loci which contribute most heavily to these estimates. Ours is the only estimate so conspicuously dominated by loci associated with glycolysis. Simultaneously, it is the only study where variability at these specific loci appears to have a readily identifiable selective value.

<u>IV. Altitude related changes in the frequencies of the common alleles at the polymorphic loci</u>. It has been argued that the neutral mutation-random genetic drift model implies that structurally different variants of any particular enzyme or protein polymorphism must be functionally equivalent. Ergo, functional dissimilarity must imply selection. Unfortunately, this argument is not rigorous. Functional dissimilarity in vitro does not necessarily imply dissimilarity in vivo. Table 2 reveals the frequencies of the common alleles at the 20 loci tested on a village-by-village basis. The rare variants enumerated in Table 1 have been excluded from these calculations. Second, no effort has been made to adjust these frequencies, and more importantly their variances, resulting from the inclusion of relatives.

Consider the monomorphic loci. Note that all of the loci known to play a prominent role in the transport or tissue utilization of oxygen fall into this category. Of those enzymes or serum proteins whose roles in oxygen transport are more uncertain, four are monomorphic. One of these, adenosine deaminase, is often polymorphic in other populations (Mourant et al., 1976). None of the six enzymes or serum proteins which presumably play no role in oxygen transport is commonly polymorphic in other populations either.

What, then, can be made of the polymorphic loci? Table 3a addresses the significance of the variability between villages without regard to altitudinal habitats, between villages within habitats, and finally between habitats. Let me hastily point out the tentative nature of this analysis. As previously stated, numerous related individuals are incorporated within the village samples, particularly

Table 2. Frequencies of the common alleles at the 20 loci under consideration by the individual's villages of residence. A locus has been designated as polymorphic, if and only if, at least two common alleles exist. The rare variants itemized in Table 1 have been excluded from these tabulations.

		Altiplano						Sierra						Coast	
		Toledo	Turco	Visviri	Caquena	Parinacota	Guallatire	Putre	Socoroma	Murmuntani	Chapaquina	Belen	Tignamar	Azapa	Lluta
ENZYMES OR SERUM PROTEINS:															
GENERALLY PRESUMED TO BE INVOLVED IN O_2 TRANSPORT															
Polymorphic															
None		--	--	--	--	--	--	--	--	--	--	--	--	--	--
Monomorphic															
The AK, G6PD, LDH, PHI, 6PGD, and Hb loci did not vary in any of the 14 villages or locales. The alleles observed are AK-1, G6PD-B, LDH-N, PHI-1, 6PDG-A, and Hb-A.															
ROLE IN O_2 TRANSPORT UNCERTAIN															
Polymorphic															
HP	1	0.802	0.651	0.640	0.740	0.672	0.705	0.615	0.614	0.711	0.727	0.725	0.626	0.641	0.585
	2	0.198	0.349	0.360	0.260	0.328	0.295	0.385	0.386	0.289	0.273	0.275	0.374	0.359	0.415
PGM_1^{-1}	1	0.830	0.833	0.804	0.725	0.775	0.921	0.808	0.659	0.895	0.654	0.788	0.789	0.773	0.797
	2	0.170	0.167	0.196	0.275	0.225	0.079	0.192	0.341	0.105	0.346	0.212	0.211	0.227	0.203
Monomorphic															
The ADA, ICD, PGM_2, and Tf loci did not vary in any of the 14 villages or locales. The alleles observed are ADA-1, ICD-1, PGM_2-1, and Tf-C.															
GENERALLY PRESUMED TO PLAY NO ROLE IN O_2 TRANSPORT															
Polymorphic															
Est	1	0.750	0.653	0.718	0.752	0.655	0.889	0.741	0.818	0.711	0.819	0.760	0.809	0.785	0.799
	2	0.250	0.347	0.282	0.248	0.345	0.111	0.259	0.182	0.289	0.181	0.240	0.191	0.215	0.201

Table 2 (Cont'd)

ACP	A	0.255	0.349	0.244	0.320	0.268	0.167	0.266	0.136	0.263	0.235	0.212	0.211	0.230	0.235
	B	0.745	0.651	0.756	0.680	0.732	0.833	0.734	0.864	0.737	0.765	0.788	0.789	0.770	0.765

Monomorphic

The PepA, PepB, PepC, MDH, Cp, Alb_8, and Alb_5 loci did not vary in any of the 14 villages or locales. The alleles observed are PepA-1, PepB-1, PepC-1, MDH-N, Cp-B, Alb_8-N, and Alb_5-N.

Table 3a. Analysis of village and habitat heterogeneity in the frequencies of the common alleles at the acid phosphatase, esterase D, haptoglobin and phosphoglucomutase-1 loci. The entries are chi-squares.

Locus	Heterogeneity among villages over habitats df=13	Heterogeneity among villages within habitats df=11	Heterogeneity among habitats df=2
Acid phosphatase	40.92**	32.91**	8.01**
Esterase D	69.49**	56.43**	13.06**
Haptoglobin	49.07**	37.54**	11.53**
Phosphoglucomutase-1	61.39**	52.87**	8.52**

**Significant at the 1% level.

in the altiplano and the sierra; the analysis set forth in Table 3a ignores this. Thus, smaller variances are ascribed to the gene frequency estimates than they in fact have. This tends, in turn, to overestimate the significance of the deviations between villages or habitats. While this underestimation of the variances seems unlikely to account for all of the significant differences indicated in Table 3a, some or many of the significant differences could be spurious. Note, however, that for all four of the polymorphic systems the variability between villages is significantly large when habitat is ignored. Observe also that habitats are uniformly different. The meaning of this is unclear; the villages within habitats also differ. Most of the intra-habitat variability appears to reside in the altiplano. In no instance are the coastal locales significantly different, and in only one case, PGM-1, are the sierran villages significantly heterogeneous. The altiplano villages, however, differ significantly one from another at every one of the four loci involved. This heterogeneity is not ascribable to a systematic difference between the Bolivian villages (Toledo and Turco) and the Chilean ones (Visviri, Caquena, Parinacota and Guallatire); however, one Chilean village, Guallatire, contributes heavily to the heterogeneity. In three of the four systems it exhibits the most discrepant gene frequencies. Guallatire is the smallest of the altiplano villages studied, and no less than 30 of the 63 individuals in this village on whom blood is available are members of 9 nuclear families. Elimination of Guallatire reduces the variability among altiplano villages to non-significance in all cases except acid phosphatase.

Examination of the gene frequencies averaged over villages within habitats (see Table 3b) certainly does not suggest a major, systematic shift in common allele frequencies with changing oxygen tension at any of these four loci. [Yen (1972) also failed to find evidence of a systematic change with altitude in acid phosphatase gene frequencies among the Quechua.] Nor does the variation between villages within habitats provide compelling support for a selectionist argument. As previously indicated, variation is greatest in the altiplano, less in the sierra, and least on the coast; but this apportionment does not take into account migration. This has been substantial in the last several decades. When migration is considered,

Table 3b. Frequencies within the three altitudinal habitats of the common alleles at the acid phosphatase, esterase D, haptoglobin and phosphoglucomutase-1 loci.

Locus	Allele	Altiplano	Sierra	Coast
Acid phosphatase	A	0.274	0.231	0.233
	B	0.726	0.769	0.767
Esterase D	1	0.733	0.778	0.792
	2	0.267	0.222	0.208
Haptoglobin	1	0.692	0.658	0.619
	2	0.308	0.342	0.381
Phosphoglucomutase-1	1	0.824	0.783	0.786
	2	0.176	0.217	0.214

the variation seen accords qualitatively with that expected under a drift model (Rothhammer, 1977). However, two of these four loci, acid phosphatase and esterase D, appear to play no role in the transport or tissue utilization of oxygen. These, therefore, would not be expected to vary systematically with changes in oxygen tension. As far as the other two loci, PGM and haptoglobin, are concerned, their roles in these processes appear minor at best. Haptoglobin may be involved through its affinity for hemoglobin; and phosphoglucomutase is the catalyst in the conversion of glucose-1-phosphate to glucose-6-phosphate, the first intermediate which can enter either the Embden-Meyerhof pathway or the pentose monophosphate shunt. Most glucose-6-phosphate arises, however, from the action of HK on glucose.

E. Whither Now?

Where, now, do these observations lead us? Clearly not as yet to an answer, but what more can be done? Several other lines of inquiry are possible; we shall describe those which we propose to pursue. First, we are currently studying ten more enzymes. These include aldolase, enolase, glyceraldehyde-3-phosphate dehydrogenase (G-3-PD), phosphofructokinase (PFK), 2,3-diphosphoglycerate mutase (2,3-DPGM), carbonic anhydrase I and II (CA I, CA II), hexokinase, phosphoglycerate kinase (PGK), and triosephosphate isomerase (TPI). As stated earlier, deficiencies of G-3-PD, PFK, HK, 2,3-DPGM, PGK and TPI are associated with non-spherocytic hemolytic anemia. A deficiency of aldolase leads to intolerance to fructose, which can be fatal if unrecognized. Again, no cases have been described among the Aymara. As yet, a deficiency of enolase has not been described. From its metabolic role, however, it is reasonable to presume that it would have serious consequences. Kendall and Tashian (1977) have recently reported an inherited deficiency of red cell CA I; all three individuals homozygous for the deficiency are normal. Their levels of CA II were not depressed, however; and it is the latter isozyme which has the higher CO_2 hydrase activity and the greater role in respiration. A deficiency of CA II has not been reported, but such a deficiency in the absence of a depression of CA I might also have little effect. The correlation in activity between these two isozymes is normally very high, suggesting one may compensate for the inactivity of the other.

Second, we intend to assess the effect of the AK, G6PD, LDH, PHI and 6PGD variants on the individuals who possess them. This evaluation will include studies at several levels of biochemical, behavioral, and physiologic function. These are ascertainment of the (1) in vitro properties of the variants as measured by Km (substrates, cofactors), Vmax, and thermal stability, and the (2) in vivo function of the variants as measured by reaction velocity at physiological levels of substrates and cofactors measured in the same individual. Reaction rates will also be determined in the presence of physiological levels of allosteric affectors, and the kinetics of activation or inhibition will be assessed when a sufficient amount of enzyme is available. (3) Levels of the metabolic intermediates of glycolysis and the pentose shunt pathways in the red cells of normal and variant individuals from the same family will be determined as will be levels of glutathione. (4) Hemoglobin function will be measured by percent O_2 saturation, level of methemoglobin, and O_2 equilibrium in whole blood. (5) Red cell age distribution will be determined by differential cell counts. (6) Venous pH, pO_2, hemoglobin content, and hematocrit will be ascertained. (7) Blood flow, and continuous (that is, 24 hour) monitoring of cardiac activity will be undertaken. (8) Oxygen consumption on a minute, hourly, and daily basis at heart rates encountered at normal activity levels will be estimated. (9) Assessment of night vision (as an indirect measure of the unloading of oxygen in the retina), and such metrics of behavior as tapping, and maze performance will be studied. (10) Finally, an effort will be made to relate these biochemical measures to our observations on health (as assessed by conventional clinical parameters), pulmonary function, and growth and development.

Third, we propose to use the perchloric acid or trichloracetic acid precipitated specimen described earlier to evaluate levels of glycolytic intermediates. Enzymatic activity may be regulated through the amounts of such intermediates, and it is this aspect of regulation which intrigues us. Evidence will also be sought to discover kinetic variants of a given enzyme that are not detectable by conventional electrophoretic techniques. Specifically, we expect to measure glucose-6-phosphate (G6P), fructose-6-phosphate (F6P), fructose-1, 6-diphosphate (FDP), dihydroxyacetone phosphate (DHAP), 3-phosphoglycerate (3PG), 2,3-diphosphoglycerate (2,3-DPG), phosphoenol pyruvate (PEP), pyruvate (PYR), and lactate (LACT). Other intermediates such as 6-phosphogluconate will be measured if practicable.

The relationships between these intermediates and the enzymes of the Embden-Meyerhof and pentose shunt pathways are complex. It is now apparent, for example, that the role of 2,3-DPG is less simple than originally conjectured (see Benesch and Benesch, 1969). It regulates, and is regulated in turn, by a variety of other products of red cell glycolysis. Brewer (1969) has shown that 2,3-DPG inhibits HK; there is also evidence that 2,3-DPG inhibits PGM, PFK, PHI, and aldolase, apparently in a competitive fashion, and G-3-PD in a non-competitive fashion (Srivastava and Beutler, 1972). It also inhibits 6-PGD, (Yoshida, 1973). Not all of the observations cited are without controversy, but the cumulative evidence attests to the central role which 2,3-DPG concentration plays in the affinity of hemoglobin for oxygen. It has been argued that the increase in 2,3-DPG with altitude is a compensatory mechanism to increase oxygen unloading to the tissues. It is also self-regulatory, since the more 2,3-DPG is bound by deoxyhemoglobin, the less are its inhibitory effects on HK and diphosphoglycerate mutase. This promotes further glycolysis which, in turn, enhances the production of 2,3-DPG (Lenfant et al., 1968; Anonymous, 1972).

Finally, we may be obliged to examine more complex selective processes than the fairly simple and straightforward notions which are implicit in our current thoughts on these issues. Consider the following illustration - one admittedly selected more for its titillating quality than its prospects of importance. The potato, a New World plant, is marvelously diverse in the altiplano; indeed, a visit to a market in La Paz or almost any other highland Bolivian community would introduce most Europeans and North Americans to several dozen previously unfamiliar varieties. These tubers have been and continue to be a major foodstuff for the Aymara; several varieties of them are used to prepare chuno negro (Sauer, 1963; Mazess and Baker, 1964). Preparation of chuno negro involves extensive freezing and thawing of the potatoes over successive days. The residual water is then stamped out, and the crushed potatoes are dried prior to storage. This preparation can be preserved easily over long periods of time. Note that the "fresh" potato is neither peeled nor cooked prior to preparation, and when chuno negro is eaten it may be chewed like bread or placed in water and boiled.

It has long been known that potatoes contain a substance, solanine, found principally in the peel and sprouts. Solanine can be toxic to humans as well as other animals. Among the symptoms of such poisoning are disturbances in respiration and cardiac activity, significant hemolysis, and other evidence of stimulation of the cholinergic system, i.e., stimulation of those nerve fibers that cause effects similar to those induced by acetylcholine (Pokrovsky, 1960). The mechanism through which these symptoms come about is apparently the inhibition of serum and brain cholinesterase. At least two enzymatically different inherited forms of serum cholinesterase are known. The three phenotypes are differentially inhibited by potato extracts, now generally thought to be solanine (Harris and Whittaker, 1959; Abbott et al., 1960). Clearly the symptoms of solanine poisoning just described would be disadvantageous, and would certainly further compromise the health of an individual already required to cope with the stress of chronic hypoxia. It seems reasonable to assume that the Aymara, who consume large quantities of potatoes, are at greater risk of solanine poisoning than individuals whose nutritional habits are different. Thus, they may have evolved a type of cholinesterase little inhibited by solanine. There is as yet no direct evidence of this, but several observations are pertinent. First, we have found no mention of the frequency with which potato toxicity is seen among Andean peoples. It seems unlikely to be high, since it has not been mentioned by several centuries of observers. Moreover, evidence of marginal or submarginal toxicity would be more interesting, for presumably such toxicity might be more widespread. Second, elsewhere we have published evidence that the effect of the mydriatic, mydriacil, upon Aymara appears to be different from that on Europeans or Mestizos. More specifically, if the "success" of instillation of two drops of mydriacil into the conjunctival sacs is judged in terms of clinically effective mydriasis in 25 to 30 minutes, fewer Aymara respond than Mestizos; and fewer of the latter respond than Europeans (Goldsmith et al., 1977). Admittedly, this is a somewhat arbitrary dichotomization of a pharmacological response which must be quantitative rather than qualitative. However, the criteria were selected before any Aymara had been examined. Dilation is generally thought to arise as a result of the anticholinergic effect of the mydriatic, i.e., as a result of the inhibition, in this instance, of the activity of acetylcholinesterase. It is conceivable, then, that the higher frequency of "non-dilators" in the Aymara is indirect testimony to a higher frequency of the gene associated with the form of serum cholinesterase which is little inhibited by solanine. This implies that solanine has a pharmacologic activity akin to an anticholinergic; whether this is so may be debatable. We are presently attempting to test this conjecture.

Acknowledgements. We gratefully acknowledge the support of the National Heart, Lung and Blood Institute through grant HL-15614, and the Multinational Genetics Program, Organization of American States, Chile.

References

Abbott, D.C., Field, K.,Johnson, E.I.: Observations on the correlation of anti-cholinesterase effect with solanine content of potatoes. The Analyst 85:375-376. (1960).

Anonymous: Editorial, Lancet ii:414-415. (1972).

Arnold, H., Hoffman, A., Englehardt, R., Lohr, G.W.: Purification and kinetic properties of glucosephosphate isomerase from human erythrocytes. In Erythrocytes, Thrombocytes, Leukocytes: Recent Advances in Membrane and Metabolic Research. (ed. E. Garlack, K. Moser, E. Deutsch, and W. Williams). Gerog Stuttgart Thieme, 1973, pp. 177-180.

Astrand, P., Rodahl, K.: Textbook of Work Physiology. New York: McGraw-Hill, 1970.

Baker, P.T.: The adaptive fitness of high altitude populations. In The Biology of High Altitude People. (ed. P.T. Baker). Cambridge: Cambridge University Press, 1977.

Barrie, S.E., Harris, P.: Effects of chronic hypoxia and dietary restriction on myocardial enzyme activities. Am. J. Phys. 231: 1308-1313. (1976).

Bellingham, A.J.: Haemoglobins with altered oxygen affinity. Br. Med. Bull. 32:234-238. (1976).

Benesch, R., Benesch, R.E.: Intracellular organic phosphates as regulators of oxygen release by hemoglobin. Nature 221:618-622. (1969).

Bertin, R., Harris, J.E., Ferrell, R.E., Schull, W.J.: The Nubians of Kom Ombo: Serum and red cell protein types. Hum. Hered. 28:66-71. (1977).

Beutler, E., Matsumoto, F., Guinto, E.: The effect of 2,3-DPG on red cell enzymes. Experientia 3:191-193. (1974).

Brewer, G.J.: Erythrocyte metabolism and function: Hexokinase inhibition by 2,3-diphosphoglycerate and interaction with ATP and Mg^{2+}. Biochim. Biophys. Acta 192:157-161. (1969).

Ferrell, R.E., Bertin, T., Barton, S., Rothhammer, F., Schull, W.J.: A multinational Andean Genetic and Health Program. IX. Gene frequencies for 20 serum protein and erythrocyte enzyme systems in the Aymara of Chile. Am. J. Hum. Genet. (submitted). (1978a).

Ferrell, R.E., Nunez, A., Bertin, T., Labarthe, D.R., Schull, W.J.: The blacks of Panama: Their genetic diversity as assessed by 15 inherited biochemical systems. Am. J. Phys. Anthro. 48:269-276. (1978b).

Ferrell, R.E., Bertin, T., Barton, S.A., Young, R., Murillo, F., Schull, W.J.: The Aymara of western Bolivia. IV. Gene frequencies for 8 blood groups and 20 serum protein and erythrocyte enzyme systems. Am. J. Hum. Genet. 30. (1978c).

Frisancho, R.: Functional adaptation to high altitude hypoxia. Science 187:313-319. (1975).

Goldsmith, R.I., Rothhammer, F., Schull, W.J.: Mydriasis and heredity. Clin. Genet. 12:128-133. (1977).

Harris, H.: The Principles of Human Biochemical Genetics, 2nd Edition. New York: American Elsevier Publ. Co., 1976.

Harris, H., Hopkinson, D.A., Robson, E.B.: The incidence of rare alleles determining electrophoretic variants: Data on 43 enzyme loci in man. Ann. Hum. Genet. 37:237-253. (1973).

Harris, H., Whittaker, M.: Differential response of human serum cholinesterase types to an inhibitor in potato. Nature 183: 1808-1809. (1959).

Harris, P.: Myocardial metabolism. In Man at High Altitude. (ed. D. Heath and D.R. Williams). Edinburgh: Churchill Livingstone, 1977.

Harris, P., Castillo, Y., Gibson, K., Heath, D., Arias-Stella, J.: Succinic and lactic dehydrogenase activity in myocardial homogenates from animals at high and low altitude. J. Mol. Cell. Cardiol. 1:189-193. (1970).

Hurtado, A.: The influence of high altitude on physiology. In High Altitude Physiology: Cardiac and Respiratory Aspects (ed. R. Porter and J. Knight). London: Churchill-Livingstone, 1971, pp. 3-13.

Kendall, A.G., Tashian, R.E.: Erythrocyte carbonic anhydrase I (CA I): Inherited deficiency in humans. Science 197:471-472. (1977).

Kirk, R.L., Keats, B., Blake, N.M., McDermid, E.J., Ala, F., Karimi, M., Nickbin, B., Shabazi, H., Kmet, J.: Genes and people in the Caspian Littoral: A population genetic study in Northern Iran. Am. J. Phys. Anthrop. 46:377-390. (1977).

Laughlin, W.S.: Origins and affinities of the first Americans. In Origins and Affinities of the First Americans (ed. W.S. Laughlin). New York: Gustav Fischer, 1977.

Lenfant, C., Torrance, J., English, E., Finch, C.A., Reynafarje, C., Ramos, J., Faura, J.: Effect of altitude on oxygen binding by hemoglobin and on organic phosphate levels. J. Clin. Invest. 47:2652-2656. (1968).

MacNeish, R.S.: Early man in the Andes. Sci. Am. 224:36-46. (1971).

MacNeish, R.S., Berger, R., Protsch, R.: Megafauna and man from Ayacucho, highland Peru. Science 168:975-977. (1970).

Mazess, R.B., Baker, P.T.: Diet of Quechua Indians living at high altitude: Nunoa, Peru. Am. J. Clin. Nurt. 15:341-351. (1964).

Mourant, A.E., Kopec, A.C., Domaniewska-Sobczak, K.: The Distribution of the Human Blood Groups and Other Polymorphisms. London: Oxford University Press, 1976.

Nagel, R.L., Bookchin, R.M.: Human hemoglobin mutants with abnormal oxygen binding. Sem. Hemato. 11:385-403. (1974).

Neel, J.V., Ferrell, R.E., Conard, R.A.: The frequency of "rare" protein variants in Marshall Islanders and other Micronesians. Am. J. Hum. Genet. 28:262-269, (1976).

Pokrovsky, A.A.: Action of alkaloids of sprouting potatoes on cholinesterase. Biokhimya 21:705-709. (1960).

Rose, I.A., O'Connel, E.L.: The role of glucose-6-phosphate in the regulation of glucose metabolism in human erythrocytes. J. Biol. Chem. 239:12-17. (1964).
Rothhammer, F.: Microdiferenciacion Genetica de una Poblacion Indigena de Habla Aymara en la Provincia de Arica. Ph.D. Dissertation, Universidad de Chile, Sede Santiago-Oriente. (1977).
Sauer, C.O.: Cultivated plants of South and Central America. In Handbook of South American Indians, Vol. VI. (ed. J.H. Steward). New York: Cooper Square Publ. Inc., 1963.
Schroter, W.: Kongentale nichtsparacytare hamolytische Anamie bie 2,3-phosphoglyceratmutase Mangel der Erythrocyten im fuhren Saulingsalter. Klin. Wschr. 43:1147-1153. (1965).
Schull, W.J., Rothhammer, F.: A multinational Andean genetic and health programme: A study of adaptation to the hypoxia of altitude. In Genetic and Nongenetic Components in Physiological Variability. (ed. J.S. Weiner). London: Society for the Study of Human Biology, Vol. 17, 1977a, pp. 139-169.
Schull, W.J., Rothhammer, F.: Analytic methods for genetic and adaptational studies. In Origins and Affinities of the First Americans. (ed. W.S. Laughlin). New York: Gustav Fisher, 1977b.
Srivastava, S.K., Beutler, E.: The effect of normal red cell constituents on the activities of the red cell enzymes. Arch. Biochem. Biophys. 148:249-255. (1972).
Tanis, R.J., Neel, J.V., Dovey, H., Morrow, M.: The genetic structure of a tribal population, the Yanomama Indians. IX. Gene frequencies for 18 serum proteins and erythrocyte enzyme systems in the Yanomama and five neighboring tribes: Nine new variants. Am. J. Hum. Genet. 25:655-676. (1973).
Tschopik, H.: The Aymara. In Handbook of South American Indians, Vol. II. (ed. J.H. Steward). New York: Cooper Square Publ. Inc., 1963.
Valentine, W.N., Tanaka, K.R.: Pyruvate kinase deficiency and other enzyme-deficiency hereditary hemolytic anemias. In The Metabolic Basis of Inherited Disease. (ed. J. Stanbury, J.J. Wyngaarden, and D.S. Frederickson). New York: McGraw-Hill, 1972, pp. 1338-1357.
Weiss, K.M., Schull, W.J.: Demography and the measurement of Darwinian fitness. Proc. IUSSP General Conference, Mexico City. Vol. 3, 1977, pp. 391-408.
Yanase, T., Hanada, M., Seita, M., Ohya, I., Ohta, Y., Imamura, T., Fujimura, T., Kawasaki, K., Yamaoka, K.: Molecular basis of morbidity - from a series of studies of hemoglobinopathies in Western Japan. Jap. J. Hum. Genet. 13:40-53. (1968).
Yen, F.: Polimorfismo a nivel molecular de los genes que determinan la fosfatasa acida del eritrocito en poblaciones humanas del altiplano Peruano. Arch. del Inst. de Biol. And. 5:1-12. (1972).
Yoshida, A.: Hemolytic anemia and G6PD deficiency. Science 179:532-537. (1973).

3 *Drosophila*

Speciation and Sexual Selection in Hawaiian *Drosophila*

H.L. CARSON

A. Introduction

How closely do the genes of an organism track the outer physical and organic properties of the ecosystem? A common assumption among evolutionary geneticists holds that this kind of natural selection acts as the most important evolutionary force at all levels from local population to the formation of species.

In a certain sense, this view is irrefutable, since no organism can afford to suffer maladjustment to its environment. What I bring into question, however, is the current tendency to ascribe all significant genetic differences within species, and between closely related species, to the effects of an all-pervading external environment.

The present paper considers the significance of another type of genetic change, namely, that required for the mutual adjustment of the sexes to what may be called the intraspecific sexual environment. An hypothesis is advanced that, in certain organisms, sexual selection demands major genetic changes within species. These adjustments may become sufficiently profound to function as isolating mechanisms if allopatric separation occurs. Indeed, in the early phases of divergence the genetics of traits of sexual and territorial behavior may far outweigh, in evolutionary importance, other types of adjustment to the ambient environment. This concept does not preclude the latter type of adaptation at a later phase in the life history of the species.

B. Background

For about 15 years, research on the Drosophilidae of Hawaii has been going forward. Recently, emphasis has shifted from the study of phylogeny (review: Carson et al., 1970) to genetical analyses of natural populations of selected species (review: Carson and Kaneshiro, 1976). Citations in the present account will be mostly confined to papers published since the latter review.

Several extraordinarily close, and apparently newly-evolved, species-pairs have been identified on the newest island, Hawaii. This paper reviews the facts concerning one such case, that of the two species, *Drosophila silvestris* and *D. heteroneura*.

Relevant to the discussion which follows, these basic facts are clear. The number of Hawaiian species, especially of the genus *Drosophila*, is very large (around 500); almost all are endemic. Most species are confined to single islands; exceptions to this occur mostly on several islands which were joined in the recent geological past. At first view, the morphological, physiological, behavioral and ecological diversity among the species is appalling. When studied more closely, however, very strong basic biological similarity is revealed, leading to the view that some of these embellishments are in a sense superficial. Indeed, the entire fauna could have evolved from a single founder specimen derived from some continent perhaps six million years ago. In any event, evolution has been conspicuous, recent, rapid and proliferative. Adaptation to many breeding substrates and to wet or dry ecosystems is also pronounced.

The six major Hawaiian islands are of volcanic origin and lie well south of the Tropic of Cancer. Each island is successively younger as one approaches the southeast from the northwest. The origin of each island in turn, over the last six million years, appears to be related to the slow passage of the Pacific crustal plate over a hypothetical "hot spot" in the earth's mantle beneath the plate. This circumstance creates an extraordinary stage for the acting out of terrestrial evolution in a sequential fashion and draws special attention of the evolutionist to the fauna of the newest island. This is the island of Hawaii ("The Big Island"), which is larger than all the rest put together; it is around 160 km long by 100 km wide. Dating studies reveal no lava flows older than 700,000 years; yet the island is formed from five large volcanos, two of which are massive shields rising to about 4300 m above sea level. Intercepting the trade winds on their northeast slopes, these volcanoes have developed exuberant tropical rainforests on mountainsides below the timberline. Leeward areas have developed a complex assortment of semi-isolated dryforest associations. The richest forests develop on soils built from volcanic ash deposited on porous lava. Nevertheless, the adaptive properties of many Hawaiian plants permit efficient colonization of new lava flows on which soil buildup occurs only slowly. Compared with the older islands to the northwest, erosion has only barely begun; the shapes of the original volcanoes are intact. All but the oldest, Kohala, lack the eroded canyons and amphitheatre-shaped valleys which characterize the older islands. Two volcanoes, Mauna Loa and its new satellite, Kilauea, are currently active.

Of particular interest to the evolutionary biologist, however, is the continuing history of the fractionation of the environment by lava flows. This phenomenon is now most easily observed on the Big Island on the slopes of the currently active volcanoes. Here, newer lava flows have surrounded and isolated patches of older forests ("kipukas"). In Hawaii, isolation occurs at every level: archipelago from continent; island from island; volcano from volcano within an island; dry side from wet side within a volcano; kipuka from kipuka on a single mountain slope.

The Hawaiian archipelago is, and indeed always has been, a highly isolated oceanic group. Accordingly, the flora and fauna of the Big Island are derived, in very large measure, from the adjacent islands to the northwest, principally the Maui island complex (Maui-Molokai-Lanai). Polytene chromosome tracing of the origin of approximately 24 large "picture-winged" Drosophila species endemic to the Big Island support this view; 20 of these have clear putative ancestors on the Maui complex.

The picture-winged Drosophilas are the most spectacular and best-known element of the Hawaiian Drosophilidae. They consist of somewhat more than 100 species of which about 36 are found on the Maui complex and 24 on the Big Island. The best judgment to date considers that no species is common to the two areas; that is, all the Big Island species are unique and have newly evolved since the formation of the island less than three quarters of a million years ago. This conclusion is based not only on a great number of clear morphological differences but in a number of instances on differences revealed by hybridization, chromosomal, electrophoretic, and behavioral techniques.

As stated earlier, all facts suggest that species endemic to the island of Hawaii are newly-formed in time. The question arises to what extent the guiding effects of environmental selection can be related to the formation of these new species and the complexes of characters that they show. To this end, genetic variation within and between certain endemic Big Island species of Drosophila has been studied, and correlations have been attempted with various ecological, behavioral and geographical parameters. One pair of these species has been chosen for detailed examination in this paper.

C. *Drosophila silvestris* and *D. heteroneura*

Morphology and relationships

The following account will center on the evolutionary biology and genetic similarity of these two species, both of which are endemic to the Big Island of Hawaii. Much of the work reported below is still in progress. *D. silvestris* is a widespread inhabitant of the luxuriant montane rainforests over most of the island where it is accompanied in many areas by the very similar but generally rarer *D. heteroneura*. Because the comparison of these two species forms the main subject of this paper, it should be emphasized at this point that these two species are very distinct morphologically. Even with the naked eye, the extraordinary mallet-shaped head of both sexes of *D. heteroneura* presents a striking difference (Figure 1).

Fig. 1. Males of *Drosophila heteroneura* (left) and *Drosophila silvestris* (right) about 12X. Photograph by K. Y. Kaneshiro.

In addition, *D. heteroneura* is somewhat smaller and much yellower; it also has distinctive pleural, abdominal, wing and leg markings. In no way do these two resemble classical "sibling" species in the sense that there is difficulty in separating them morphologically.

The two species belong to a distinct chromosomal subdivision of the *planitibia* subgroup of the picture-winged Drosophila; their closest relatives are *D. planitibia* of Maui and *D. differens* of Molokai. All four are morphologically distinct.

Ecology and Distribution

D. silvestris is generally confined to a fairly narrow altitudinal zone between 900 and 1500 m on high-rainfall slopes. Although it will breed on the fermenting parts (mostly small branches) of a number of endemic host plants, it is characteristically abundant only in forests which have numerous arborescent lobeliads in the understory, especially *Clermontia hawaiiensis* and its relatives. This plant is an endangered species which is only locally common, and is absent from large areas in the mesic windward forests. This absence appears to be mostly a natural circumstance, not related to any destruction of *Clermontia* by man. Populations of *D. silvestris* are restricted accordingly by the distribution and abundance of this primary host plant.

D. silvestris may be easily captured with the use of decaying mushroom and fermenting banana baits. In most places, 3-6 specimens per man-hour can be collected. Occasionally a rich natural accumulation can be found, and flies can be individually collected at about twice the above rate. Firm population censuses have not been made, but there is little question that the population sizes are small compared to the widely-studied continental species. The collector must be satisfied with sampling techniques and results comparable to those for many terrestrial vertebrates.

D. silvestris is more widespread than *D. heteroneura*, and in many areas it occurs alone at the higher altitudes. Furthermore, it appears to occur alone on the Kohala Volcano, the geologically oldest area on the island. No place has been found where *D. heteroneura* occurs allopatrically, and in only one area has it been found to outnumber *D. silvestris*. Furthermore, this situation may have been a temporary one.

Although not invariably so, *D. heteroneura* appears to favor areas of forest on newer lava flows where the forest is more open. It appears to have some latitude of host plants for ovipositional sites, but, like *D. silvestris*, it breeds well on *Clermontia hawaiiensis*. Several carefully documented cases of the eclosion of both species from the same decaying *Clermontia* branch are known. The two species are in intimate association in nature, and frequent encounters between individuals of the different species can be observed.

Territorial behavior

Both species display elaborate territorial and sexual behavior (Spieth, 1966). Males, usually predominantly of one species, accumulate in spaced concentrations, called leks. In *D. silvestris*, these characteristically occur on the various fronds of a single large tree fern. Each male defends a small territory on a rachis within the larger lek site, aggressively approaching any intruder. The flies apparently have high visual acuity and display much wing-waving before bouts of contact behavior ensue. Females appear to be dispersed more evenly in the forest. Courtship and mating is a highly complex process which appears to occur only at lek sites. The females move to the leks singly; copulations are only rarely observed.

Isolation

Both species are long-lived (3-6 months) and survive well in the laboratory. When presented with a suitable substrate, females will readily oviposit, producing laboratory stocks. This has permitted the study of sexual isolation. Virgin males and females are collected from laboratory strains at eclosion and are aged, the two sexes separately, on standard medium, until they are about four weeks old. Then a single female of each of the two species is placed together with a male of either species, (Kaneshiro, 1976). Such a small experimental unit is required because the males are extraordinarily aggressive towards one another. This trio of flies (two females and one male) is kept together on standard medium until one of the females is inseminated. When applied to *D. silvestris* and *D. heteroneura* in both reciprocals, this method reveals a very strong ethological isolation between the species. Using the Charles-Stalker index, the isolation in both directions is the same (0.92 on a scale in which 1.0 denotes only homogamic matings). The above studies were done with laboratory strains, each derived from a single wild female, taken from areas about 5 km apart on the slopes near Kilauea volcano. Preliminary indications are, however, that flies derived from an area of close sympatry are behaviorally isolated to about the same degree. These findings reinforce the morphological indication that the two entities under consideration do indeed behave as two isolated sympatric species.

Inversion polymorphism

Females of both species captured in nature were placed individually into culture tubes in the laboratory. F_1 larvae were obtained, and cytological analyses of metaphase and polytene chromosomes were carried out on these larvae. Although the metaphase chromosomes show no consistent differences, considerable chromosomal polymorphism is revealed in the polytene chromosomes, especially in *D. silvestris*.

D. heteroneura is sequentially monomorphic in the polytene chromosomes except for the presence of a single inversion, distally located in chromosome 3 (3m). Band-order comparisons show that, except for the 3m section, this species is homokaryotypic for the same gene orders in all chromosomes as the Standard for the subgroup, *D. planitibia*. Because *D. heteroneura* shares such identical arrangements with *D. planitibia*, including the standard chromosome 3, the two species may be referred to as homosequential. *D. silvestris* may also be designated as homosequential with the other two, although this species displays seven widespread polymorphic inversions. An eighth has been found only once. In a number of populations, some of the inverted sections occur in high frequency. Accordingly, the Standard polytene karyotype (i.e. Standard order for all chromosomes) is very rare in populations of *D. silvestris*. The geographical distribution and frequencies of the seven widespread inversions of *D. silvestris* are shown in Table 1 (selected data from Craddock and Johnson, ms).

Table 1. Population frequencies of the seven polymorphic inversions of *D. silvestris*. (Adapted from Craddock and Johnson, ms)

Population	No. of chromos. sampled	Inversion frequencies						
		2o	3m	3r	$4k^2$	4t	$4l^2$	$4m^2$
Kilauea Forest Reserve	436	0.028	0.324	0	0.511	0.782	0.179	0.016
Keahou Ranch	86	0	0.616	0.023	0.698	0.953	0.012	0.012
Kipuka at 4400'	51	0	0.941	0.529	0.863	0.640	0.039	0.255
Mawae Kipuka 9	172	0.046	0.552	0.128	0.395	0.709	0.256	0.087
Mawae Kipuka 14	166	0.121	0.564	0.048	0.404	0.843	0.018	0.066
Puu Laalaau, Kohala	44	0.186	0.909	0.545	0.705	0.773	0	0.182
Moanuiahea, Hualalai	70	0	0.838	0.074	0.571	0.800	0	0
Kipuka Pahipa	58	0	0.466	0	0.241	0.569	0.362	0

The geographical localities listed are from widely dispersed areas of the Big Island; a map showing these sites may be found in Carson and Johnson (1975). Several important facts are obvious. *D. silvestris* shows striking local variation in inversion frequencies. In fact, there are statistically significant differences between all of the populations, including the two Mawae kipukas (Table 1) which are separated by an 1855 lava flow which is only about 1.5 km wide. These data suggest that *D. silvestris* has low vagility, but direct data on this point are lacking. The genetic information from inversions suggests that *D. silvestris* is "currently an aggregation of discrete, isolated gene pools of restricted size" (Craddock and Johnson, ms).

In order to facilitate comparison between the two species, three sites were selected where both species occur sympatrically at substantial frequencies. Information on the inversion frequencies of both species were obtained (Table 2); some of the samples have been supplemented by data given by Craddock and Johnson for the exact same sites. A map showing the locations of these three sites is given by Sene and Carson (1977).

The data of Table 2 demonstrate the strong interspecific differences in chromosomal polymorphism, especially in chromosome 4. For example, both $4k^2$ and 4t inversions are at a high frequency in *D. silvestris* but are wholly absent in *D. heteroneura*. Inversion 3m is polymorphic in both species but at each site the frequency is significantly lower ($p < .01$) in *D. heteroneura*.

Accordingly, it may be concluded that the inversion patterns, like the morphology, distribution, and behavior, indicate that two specifically distinct gene pools exist in nature. Nevertheless, in the light of this conclusion, I will now present information which appears, at least on the surface, to be in conflict with the above conclusion.

Table 2. Inversion frequencies of the two species compared at three sympatric sites.

Site and species	No. of chroms. sampled	Inversion frequencies				
		3m	3r	$4k^2$	4t	$4l^2$
Kahuku						
D. silvestris	178	0.910	0	0.787	0.815	0
D. heteroneura	96	0.469	0	0	0	0
Olaa						
D. silvestris	224	0.955	0.004	0.817	0.955	0.006
D. heteroneura	56	0.679	0	0	0	0
Pauahi						
D. silvestris	194	0.706	0.010	0.655	0.799	0
D. heteroneura	158	0.576	0	0	0	0

Allozymic comparisons

Preliminary studies of 12 loci encoding for enzymes (Johnson et al., 1975; Craddock and Johnson, ms) suggested that *D. silvestris* and *D. heteroneura* were extraordinarily similar; that is, that the coefficient of similarity was close to 0.9. In view of the other differences between the two entities given in the above sections, this result was surprising. Such a level of similarity, using similar techniques, had been described as characteristic of local populations of the *same* species (Ayala et al., 1974). Accordingly, Sene and Carson (1977) have investigated the allozymic similarity of *D. silvestris* and *D. heteroneura* further by studying 25 loci from three sympatric sites (Kahuku, Olaa, Pauahi, see above).

The data are given in schematic form in Figure 2, and the similarity coefficients summarized in Table 3. An occasional allozyme shows near-fixation within a single population of one of the species, but this is not reflected by similar changes in other populations. Accordingly, members of these two species cannot be diagnosed electrophoretically (e.g. Ayala and Powell, 1972). Also, there is no consistent geographical or altitudinal pattern within either species.

Table 3. Summary of protein differences between *D. heteroneura* and *D. silvestris* (from Sene and Carson, 1977)

	Within *silvestris*	Within *heteroneura*	Between *silvestris* and *heteroneura*
Mean Nei I	0.961 ± 0.01	0.949 ± 0.02	0.939 ± 0.01
Interpopulation differences in gene frequency (P = < 0.5)	7.3 ± 1.5	9.3 ± 1.5	10.3 ± 0.9

Hybridization in the laboratory

As described in the above section on isolation, heterogamic matings in the laboratory are rarely observed; and ethological isolation between the species is very strong. Nevertheless, rare reciprocal matings have occurred. Females so inseminated were separately placed in culture vials and permitted to oviposit. In most cases abundant F_1 larvae were produced. Apparently normal pupation and adult eclosion followed. Sex ratios at eclosion were normal; both sexes were vigorous and healthy. When tested, male and female hybrids of the reciprocals proved to be highly fertile in both F_2's and backcrosses. Detailed work on the fitness (survival and fertility) of F_1, and F_2 hybrids is still in progress, but no dramatic breakdown appears to be occurring (J. N. Ahearn, personal communication).

Inheritance of morphological species differences

The obtaining of fertile hybrids of both sexes between specific entities is unusual in the animal kingdom, although such hybrids are often found between plant species. Their existence provides an

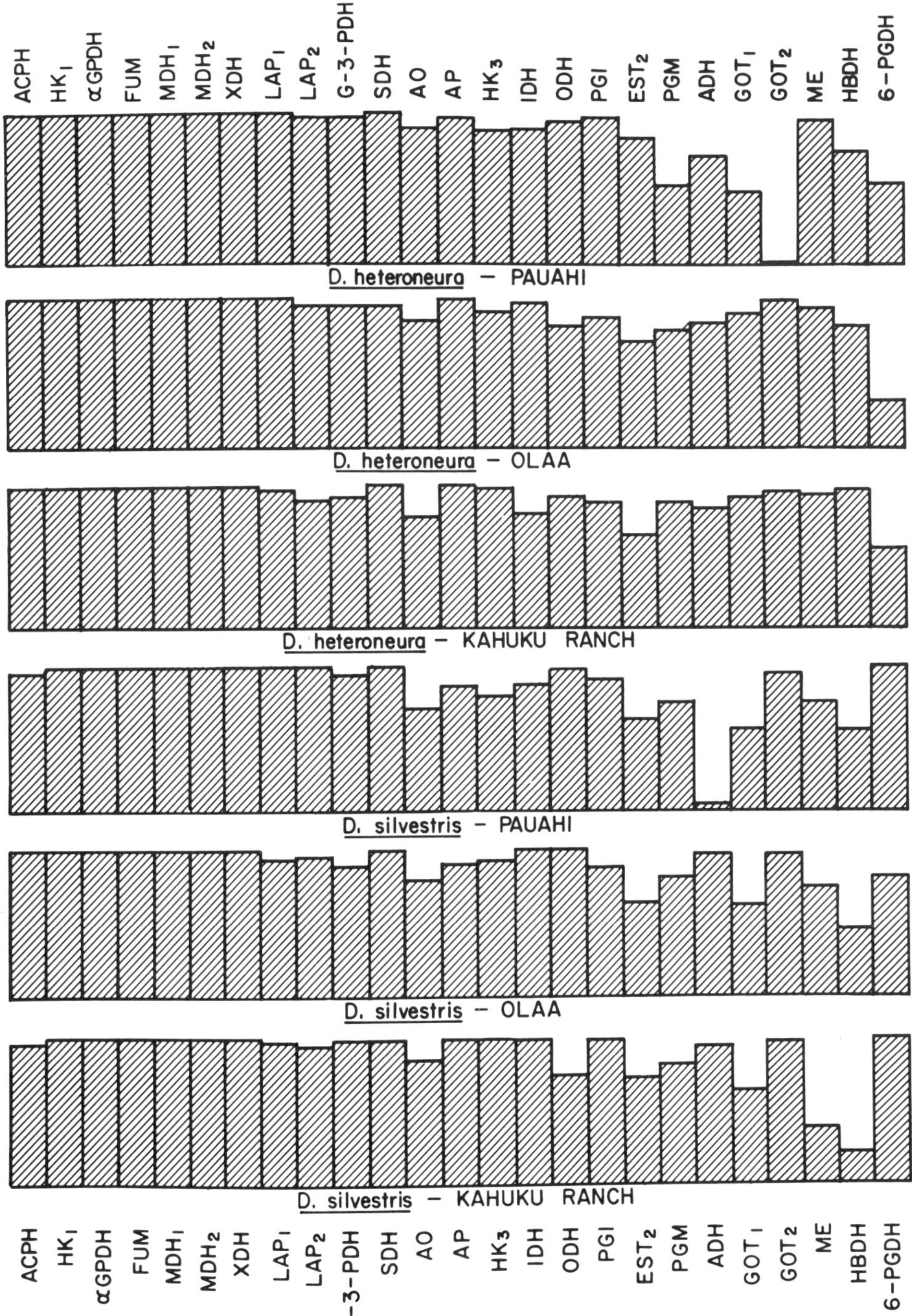

Fig. 2. Frequency of allele 1.0 at each of 25 allozyme loci in two *Drosophila* species at each of three localities.

opportunity for estimating the inheritance of the morphological attributes on which the species descriptions are based. Between these two species, many such differences may be observed, including head shape, face and antennal color, mesonotal and pleural color pattern, abdominal color pattern, leg color and the pattern of spots on the wings. Kambysellis (1974) found that the pattern of the chorion of the egg of the two species is different, and he has shown hybrid intermediacy.

Val (1976, 1977) has made a genetic study of the inheritance of these differences, especially the shape of the head, which represents the most striking feature by which the species differ. F_1 hybrids show head shapes which are intermediate between the parents. There is also a maternal effect; that is, the head shape of F_1's in the two reciprocals inclines somewhat towards the female parent. Study of F_2's reveals a broad and profound segregational dispersion, suggesting polygenic variation, involving both sex-linked and autosomal loci. Using statistical methods devised by Templeton (1977), head shape appears to depend on about 8 independently segregating genes, one of which is sex-linked. Inheritance of wing, pleural and face color patterns appears to require invocation of about six to ten more loci. Not including evidence for independent genetic control of abdominal pattern and egg chorion differences, the number of genes observed in this way amounts to differences between the two species of about 14-18 loci.

This number should not be taken too literally. Based on other genetic hypotheses, gene interactions could produce the same phenotypes from fewer differences. However, the order of magnitude appears to be correct. Accordingly, it seems clear that substantial genetic difference exists between these species, despite the electrophoretic similarities and the interfertilities which have been previously described.

Hybrids in nature

Since the early years of this work on *D. silvestris* and *D. heteroneura* thousands of specimens have been taken in collections. These specimens may easily be sorted into two groups with no intermediate forms resembling laboratory hybrids. However, in December of 1974, at Kahuku Ranch between 1000 and 1200 m on the slope of Mauna Loa, Kaneshiro captured the first intermediate to be observed, a male exactly conforming to the phenotype of an F_1 laboratory hybrid between a *D. silvestris* ♀ and a *D. heteroneura* ♂ (Kaneshiro and Val, 1977).

Since this discovery, collections in nature have been intensified in areas where the two species are sympatric. At Kahuku, between one and two percent of the wild flies show some evidence of hybrid origin. At other places, no evidence of hybridization has been found. As expected from laboratory findings, the Kahuku hybrids are fully fertile; in fact, in several instances a wild male has been successfully backcrossed to females of both parent species. A metrical analysis of the Kahuku natural populations to see to what extent one or both species are affected by introgression is currently under way. Preliminary data, based on knowledge of the phenotypes of laboratory hybrids, indicate that the effect may be small.

The possibility that the close electrophoretic similarity may be due to hybridization and introgression has not escaped the attention of our group. Several facts, however, seem to refute this view. First, there is the nearly complete chromosomal integrity of *silvestris;* at the three sympatric areas $4k^2$ and 4t inversions retain very high frequencies, especially at Olaa. Second, the two species are chromosomally most similar at Pauahi, where no hybrids have been found (Table 2). Third, by the same argument, electrophoretic similarity is greater at Olaa, also a site where no hybrids have been found, than at the site of hybridization at Kahuku (Sene and Carson, 1977). These arguments, however, apply only to the current situation; present conditions could have been affected by events in the recent or remote past.

D. Discussion and Conclusions

Sharp morphological differences and strong sexual isolation at first suggest the conclusion that *D. silvestris* and *heteroneura* are a pair of quite conventional sympatric species. Thus, females captured in nature, with only a few exceptions, produce progeny wholly conforming to the character-syndrome of the mother's species. Genetic analysis of the morphological differences between the species indicates that these features may be controlled by between 14 and 18 polygenic loci. Wild females showing the *D. silvestris* phenotype produce laboratory F_1's which manifest from one to six polymorphic inversions; none of these is associated with *D. heteroneura* characters. The two differ in geographical distribution; *silvestris* occurs alone at higher altitudes and in the Kohala mountains. Certain evidence, however, requires that the above simple characterization be closely scrutinized. In the first place, the species are extraordinarily close ecologically. In sympatric areas, the two entities breed on the same substrates, and males establish their territories in very similar sites. Both sexes of both species meet regularly in nature. Electrophoretic analysis at 25 loci reveals that the biochemical polymorphisms of the two species are very similar, on the order of similarity expected among local populations of a single species. Because strong genetically-based morphological and behavioral differences exist, it is concluded that the biochemical distance considerably underestimates the true genetic distance between the entities.

An even more striking fact, which causes some critics to question the designation of these populations as species, is the interfertility of members of the two groups. Within the constraints of the experimental analyses made so far, fertility of F_1, F_2 and backcross hybrids of both reciprocals seems quasi-normal if not fully normal. Furthermore, when sexual isolation is weakened in nature in some unknown manner, there is apparently no insurmountable barrier to the formation of natural hybrids. Nevertheless, preliminary observations indicate that this hybridization process, in the one area where it has been observed, has not weakened or diluted the morphological and cytological integrity of the population of either species. Although they are fully vigorous and fertile under laboratory conditions, some reduced fitness of these hybrids in nature is implied.

The situation regarding *D. silvestris* and *D. heteroneura* is best understood when it is recalled that they are endemic to a geologically new island and are inhabitants of the still newer rainforest ecosystem of this island. The conclusion that these species are very new in the historical sense (neospecies) is inescapable. This circumstance may be invoked to explain the great electrophoretic similarity. Much attention has been recently given to the hypothesis that this kind of genetic difference is accumulated at a constant rate within the constraints of a particular genetic system and evolutionary pattern. Thus, these differences may represent a sort of clock-like mechanism which can be used to measure the time since the two presently-existing lines diverged from their common ancestor. In the case of the two species under consideration here, the time of divergence is estimated to be about 0.2 million years (Carson, 1976).

Newness may also explain the lack of clear evidence for post-mating isolation (hybrid sterility or breakdown). If the species arose in allopatry through a founder effect and genetic reorganization, these events are not under any obligation to be accompanied by post-mating sterility or recombinational breakdown. Furthermore, in the presence of a strong premating isolation, two such entities, after they have become sympatric, need not develop such post-mating effects, as will now be argued.

This consideration leads to an important conclusion pertaining to the nature of premating isolation in these species. The enormous proliferation of Drosophila species in Hawaii seems to have a stochastic element, promoted in large measure by the founder effect. Although pronounced adaptive changes occur in some instances, in others, as in the present case, such changes appear to be of minor significance in the speciation process.

What sort of process, then, provides for the allopatric isolation of the two gene pools so that the two entities may remain separate after they have come together again? It is tempting to suggest that this is related to the extraordinary and complicated territorial and mating behavior displayed by these species.

Within a small semisolated population, in which the variability of the gene pool may have been reduced by founder effects, a reorganization of the genetic basis of the previous behavioral syndrome may occur. This has been suggested in a recent paper by Kaneshiro (1976). In these flies it is likely that sexual selection favors the buildup, over a period of time, of a complex genetically-based signal and response recognition system between male and female. Paterson (1976) has called this the "mate recognition system". Evidence indicates that the female to a large degree controls whether mating will or will not occur. Kaneshiro shows that females of "ancestral" species from geologically older islands frequently refuse matings with males of derived species, but that the reverse is not true.

Accordingly, it may be suggested that populations of organisms with complex mating behaviors may often undergo a drastic reorganization of these behaviors when a new allopatric population is founded. Random drift resulting from inbreeding and small population size may force a renewed sexual selection which reorganizes the behavioral coadaptation of male and female. In Hawaiian Drosophila, this may result in the rather rapid evolution of a new behavioral syndrome, including morphological characters which enhance mating success.

Morphological embellishments, however, are not a necessary result of the sexual selection system which promotes mate recognition. For example, the discriminatory power of the female appears to consist largely of cryptic behavioral traits, without morphological elaboration. Even in the male, there is evidence that a behavioral pattern may develop fully without being noticeably reflected in morphology. Under continued sexual selection, however, a morphological embellishment may evolve. For example, males of all 14 species of the *adiastola* group of Hawaiian *Drosophila* have, as part of their display, a ritual in which the male stands directly in front of the female, in a close head-to-head position. As excitation increases the male holds this position and then curves his abdomen upward, in the manner of a scorpion, until it is nearly touching both his own and the female's head. The abdomen is then actively vibrated. Except for some muscular change, no obvious morphological alteration has occurred at the tip of the abdomen in 13 out of the 14 species. In *D. clavisetae*, however, the end of the male's abdomen has acquired a brush of long clavate hairs, which sweep over the female's head. This species, found only on East Maui, is, according to its chromosomal relationships, phylogenetically very advanced. In this case, the behavior clearly evolved first and was embellished later with the morphological attribute.

Behavioral evolution can therefore proceed in a complex fashion even while morphological similarity is maintained. Indeed Nevo (1969) has suggested that such behavioral evolution may underlie the cryptic chromosomal species found in the mole rat, *Spalax*. It is striking that in this case, and in the similar case of the pocket gophers *Thomomys* (Nevo and Shaw, 1972; Nevo et al., 1974), biochemical differentiation has been very slight, just as in the case of *D. silvestris* and *D. heteroneura*. Cryptic species continue to be found in other mammals (e.g. Yonenaga et al., 1975), and it is possible that premating behavioral isolation is contributing to many such cases.

One may visualize *D. silvestris* and *D. heteroneura* as arising very recently by the founder effect, allopatrically, from a common ancestor (or ancestors) on the Maui Island complex. During the allopatric phase, each new population could have built its own unique mate recognition system. When the two met, the premating differences served to reproductively isolate the two populations. Interspecific crosses, when they did occur, produced individuals which were behaviorally relegated to the extremes of the normal distribution of sexual behaviors in the parent populations. These individuals thus became the victims of a stabilizing sexual selection which excludes individuals showing extreme behaviors from reproduction (see Milkman, 1973). The resulting populations thus manifest themselves as *ethospecies*, or species largely differentiated by two quite different premating behavioral syndromes.

As Val (1977) has shown, the genetic changes which cause this result are likely to be numerous, but their fixation by sexual selection need not require correlated biochemical differentiation. Thus, the minor electrophoretic distance probably reflects only the passage of time, as was argued earlier. The proposal outlined above does not require the buildup of isolation by schemes involving postmating disturbances such as hybrid sterility or segregational breakdown. "Sterility" results only in the sense that hybrids are incapable of successfully competing in the complex premating sexual contests.

It may be concluded that the species analyzed here are newly formed ethospecies kept separate by premating isolation. Their newness causes them not to display conventional hybrid sterilities or ecological or biochemical differentiation. These are likely to be acquired by such species only at some later period in their life history. Indeed, hybrid sterility and breakdown may be wholly incidental accompaniments of differential adaptation.

This ethological concept may perhaps be best invoked for animals which have elaborate mating behaviors. On the other hand, it is indeed possible that the same idea might be applied to the more passive sexual coadaptations which characterize the plant kingdom, as has been pointed out by Janzen (1977).

E. Summary

Recent studies of *Drosophila silvestris* and *D. heteroneura* from the geologically new island of Hawaii are reviewed. Although the two are broadly sympatric and are morphologically, behaviorally and cytologically distinct, they are remarkably close ecologically and electrophoretically. Furthermore, laboratory hybrids display what appears to be full fertility. One sympatric area shows about 1-2% natural hybrids, but there is no evidence that the genetic integrity of either species is breaking down in this area. It is concluded that the species were recently formed as allopatric entities, arising by the founder effect. Under the forced genetic reorganization occurring after the founder event, sexual selection has apparently produced two distinct mate recognition systems. These have been reinforced after sympatry occurred wholly at the premating level. The electrophoretic similarity is due to the recency of all these events; the allozymes have not participated in any important genetic changes related to speciation; nor have infertilities arisen as byproducts of adaptive or behavioral change. These, and possibly many other *Drosophila*, birds, and mammals may be looked upon as *ethospecies*, a term which emphasizes that the primary genetic change in speciation is the build-up of a new coadapted syndrome of mating behavior having an extensive genetic basis.

Acknowledgements. This paper was written when I was a guest at the Instituto de Biociências, Universidade de Sao Paulo, Brasil. I am grateful to Professors C. Pavan and A.B. de Cunha for many courtesies. I thank Dr. O. Frota-Pessoa for stimulating discussions on sexual selection. Experimental work was supported by BMS 74-22532 from the National Science Foundation.

References

Ayala, F.J., Powell, J.R.: Allozymes as diagnostic characters of sibling species of *Drosophila*. Proc. Natl. Acad. Sci. U.S. 69:1094-1096. (1972).

Ayala, F.J., Tracey, M.L., Hedgecock, D., Richmond, R.C.: Genetic differentiation during the speciation process in *Drosophila*. Evolution 28:576-592. (1974).

Carson, H.L.: Inference of the time of origin of some *Drosophila* species. Nature 259:395-396. (1976).

Carson, H.L., Hardy, D.E., Spieth, H.T., Stone, W.S.: The evolutionary biology of the Hawaiian Drosophilidae. In Essays in Evolution and Genetics in Honor of Theodosius Dobzhansky. (ed. M.K. Hecht, W.C. Steere). New York: Appleton-Century-Crofts, 1970, pp. 437-543.

Carson, H.L., Johnson, W.E.: Genetic variation in Hawaiian *Drosophila* I. Chromosome and allozyme polymorphism in *D. setosimentum* and *D. ochrobasis* from the island of Hawaii. Evolution 29:11-23. (1975).

Carson, H.L., Kaneshiro, K.Y.: *Drosophila* of Hawaii: Systematics and ecological genetics. Ann. Rev. Ecol. Syst. 7:311-345. (1976).

Craddock, E.M., Johnson, W.E.: Genetic variation in Hawaiian *Drosophila* V. Chromosomal and allozymic diversity in *Drosophila silvestris* and its homosequential species. (MS).

Janzen, D.H.: A note on optimal mate selection by plants. Amer. Nat. 111:365-371. (1977).

Johnson, W.E., Carson, H.L., Kaneshiro, K.Y., Steiner, W.W.M., Cooper, M.M.: Genetic variation in Hawaiian *Drosophila* II. Allozymic differentiation in the *D. planitibia* subgroup. In Isozymes IV. Genetics and Evolution (ed. C.L. Markert). New York: Academic Press, 1975, pp. 563-584.

Kambysellis, M.P.: Ultrastructure of the chorion in very closely related *Drosophila* species endemic to Hawaii. Syst. Zool. 23: 507-512. (1974).

Kaneshiro, K.Y.: Ethological isolation and phylogeny in the *planitibia* subgroup of Hawaiian *Drosophila*. Evolution 30:740-745. (1976).

Kaneshiro, K.Y., Val, F.C.: Natural hybridization between a sympatric pair of Hawaiian *Drosophila*. Amer. Nat. 111:897-902. (1977).

Milkman, R.: A competitive selection model. Genetics 74:727-732. (1973).

Nevo, E.: Mole rat *Spalax ehrenbergi:* Mating behavior and its evolutionary significance. Science 163:484-486. (1972).

Nevo, E., Kim, Y.J., Shaw, C.R., Thaeler, C.S.: Genetic variation, selection and speciation in *Thomomys talpoides* pocket gophers. Evolution 28:1-23. (1974).

Paterson, H.E.: In Symposium address. 16th Int. Congress Entomology. Washington D.C. (1976).

Sene, F.M., Carson, H.L.: Genetic variation in Hawaiian *Drosophila* IV. Close allozymic similarity between *D. silvestris* and *D. heteroneura* from the island of Hawaii. Genetics 86:187-198. (1977).

Spieth, H.T.: Courtship behavior of endemic Hawaiian *Drosophila*. Univ. Texas Publ. 6615:245-313. (1966).

Templeton, A.R.: Analysis of head shape differences between two interfertile species of *Drosophila*. Evolution 31:630-641. (1977).

Val, F.C.: Genetics of morphological differences between two interfertile species of *Drosophila*. Genetics 83:s78 (abstract) (1976).

Val, F.C.: Genetic analysis of the morphological differences between two interfertile species of Hawaiian *Drosophila*. Evolution 31: 611-629. (1977).

Yonenaga, Y., Kasahara, S., Almeida, E.J.C., Peracchi, A.L.: Chromosomal banding patterns in *Akodon arviculoides* (2n = 14) *Akodon* sp. (2n = 24 and 25) and two male hybrids with 19 chromosomes. Cytogenetics 15:388-399. (1975).

Ecology and Genetics of Sonoran Desert *Drosophila*

W.B. Heed

A. Introduction

It is axiomatic that ecological geneticists be intimately acquainted with the habits and habitats of the organisms with which they are concerned. Knowledge of even a few of the environmental cues that initiate a behavioral response can be fundamental for an eventual understanding of gene-environment interaction. However, ecological genetics is derived from fields as profound as phylogenetic histories on the one hand and biogeographic origins on the other.

Figure 1 illustrates the model of study we follow with our work on the *Drosophila* of the Sonoran Desert. The model assumes that biogeography and phylogeny are related only indirectly through ecological genetics. The areas of genetics and ecology lie between the base and apex of the triangle. The former includes a formal description of gene products and chromosome structure, their interactions and frequencies. The latter incorporates a formal description of niche and habitat, which determine the demographic properties of the species involved.

Our approach to ecological genetics is to uncover the forces maintaining each species in its habitat and to examine the genetic elements that respond to and guide them. This relationship can be elucidated readily for phytophagous insects by the analysis of host plant selection.

This report summarizes the host plant preferences, and much of what is known of the genetics and ecology of the four species of *Drosophila* endemic to the Sonoran Desert. Detailed information is presented on *D. mojavensis*, because this endemic species holds much promise for future investigations.

B. Host Plant Selection

Table 1 summarizes host plant utilization of the four species of *Drosophila* endemic to the Sonoran Desert and adjacent areas. These data were obtained over a 15-year period, and are based on approximately 68,000 flies reared from field collected cacti. The host plants are highly specialized columnar cacti of the tribe Pachycereae (*sensu* Gibson and Horak, 1977), whose main distributions are also within the Sonoran Desert. The closest relatives of each of the four species of *Drosophila* and each of the five principal host species of cacti, aside from minor offshoots, are found outside the Sonoran Desert. Therefore, for the *Drosophila*, there has been speciation into the desert but none within it. Thus, any unusual patterns of similarity among these four species of *Drosophila* or among the five principal species of host plants must be viewed as the result of parallel evolution.

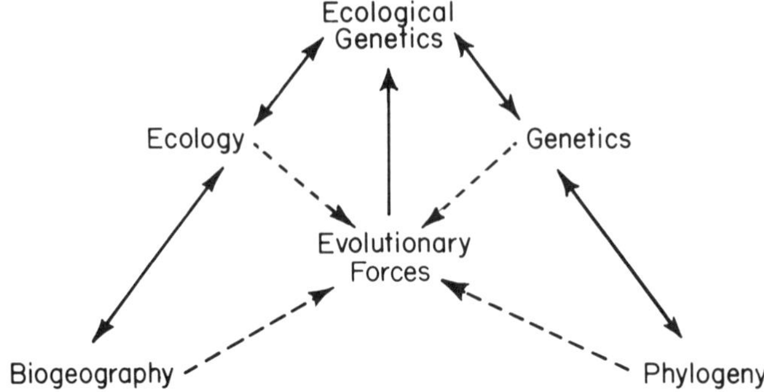

Figure 1. Relationship of ecological genetics to related disciplines (descriptive, solid lines) and the contribution of each one to the forces acting upon populations (dynamics, dotted lines).

Wiggins (1960) has discussed the disparate origin of the desert flora of Baja California, noting that approximately 40 families of plants are represented by a single genus and often by a single species. This phenomenon clearly relates to the extreme environmental stress which exists in the western portion of the Sonoran Desert.

The radiation of the platyopuntia group of the subgenus *Opuntia* is responsible for the main radiation of the repleta species group of *Drosophila* in the semi-arid parts of Mexico (Patterson and Stone, 1952). These plants are conspicuously absent in the more arid portions of the Sonoran Desert, due to their sensitivity to prolonged drought (Shreve and Wiggins, 1964). Thus, the desert of Baja California regularly supports only the four *Drosophila* species listed in Table 1. By contrast, approximately 15 additional species of *Drosophila*, several of which are closely related, and none of which is endemic to the Sonoran Desert (Heed, unpublished), are found in areas north, south, and east of this region and where the platyopuntias are utilized as host plants by many of these species.

Table 1 shows that a high degree of host plant specificity obtains within this area and with no species overlap. The chief agents which control this specificity are the toxicity, and possible nutritional deficiency, of several of the host plants to non-resident species and to competitive exclusion by the resident species in the host plants that are not toxic (Fellows and Heed, 1972). The moist soil habitat for the larvae of *D. mettleri* illustrates a niche separation rarely seen in the genus *Drosophila* (Heed, 1977). The only other reported example concerns a pair of Hawaiian *Drosophila* which also live in very arid conditions (Kaneshiro et al., 1973).

There are exceptions to the absence of species overlap in larval habitats, but they occur outside of the extremely dry peninsula (Table 1). In northern Sonora and southern Arizona *Drosophila nigrospiracula* sometimes shares its saguaro habitat with *D. mojavensis* and *D. mettleri* (Fellows and Heed, 1972; R. L. Mangan, personal communication). *Drosophila nigrospiracula* also regularly shares the

Table 1. The major host plants of the four endemic species of *Drosophila* of the Sonoran Desert.

Host plant	Abundance on peninsula	Drosophila species overlap	Abundance on mainland	Drosophila species overlap	Resident species
Lophocereus schottii (senita)	+++	No	+++	No	*D. pachea*[d]
Machaerocereus gummosus (agria)	+++	No	+[a]	No	*D. mojavensis*[e]
Stenocereus thurberi (organ pipe)	+++[b]	No	+++	No	*D. mojavensis*
Carnegiea gigantea (saguaro)	-[c]	-	+++	Yes	*D. nigrospiracula*[e]
Pachycereus pringlei (cardón)	+++	No	+[a]	Yes	*D. nigrospiracula*
Saguaro soil	-[c]	-	+++	No	*D. mettleri*[e]
Cardón soil	+++	No	+[a]	No	*D. mettleri*

[a]Host plant rare in mainland compared to Baja California but regularly used when present.

[b]Host plant common in Baja California but rarely utilized.

[c]Host plant absent in Baja California.

[d]Nannoptera species group.

[e]Repleta species group.

cardón habitat with *D. spenceri*, and with other species to a lesser degree, in southern Sonora and the Cape region of Baja California (Fellows and Heed, 1972).

Table 1 also demonstrates a shift in host plants by three of the four species. *D. mojavensis* shifts from agria in Baja California to organ pipe in Sonora except for a 30 kilometer strip along the coast of the Gulf of California from Desemboque del Rio San Ignacio to Punta Chueca where larvae of *D. mojavensis* regularly inhabitat agria. Even though organ pipe is an abundant plant on the peninsula and in the Desemboque region, it is usually not utilized by *D. mojavensis*, possibly because necrotic tissue is rarely encountered in this plant. Evolutionarily, agria and organ pipe have developed from the same phylad although they have closer relatives living elsewhere (Gibson and Horak, 1977). Their basic chemistry is not markedly different (H. W. Kircher, personal communication). Even so, the shift from one host to another by *D. mojavensis* is correlated with changes in gene and chromosomal frequencies and mating propensity.

Another shift in host plant preference occurs with *D. nigrospiracula* and *D. mettleri*. On the mainland, both cardón and saguaro are utilized as hosts, however cardón is mostly limited to the coastal region. *D. nigrospiracula* inhabitats necrotic tissue, while *D. mettleri* utilizes the soil moistened by the rots. In Baja California saguaro does not exist, and cardón is the only host. Since cardón and saguaro belong to the subtribe Pachycereinae, they are at least distantly related (Gibson and Horak, 1977).

D. nigrospiracula has been the subject of several studies on migration and genetics. Cooper (1964) found no inversion differences in hybrids from five localities in Arizona and Sonora and one from San Felipe, Baja California. Studying an esterase locus, Sluss (1975) found very close genetic similarities between 19 localities extending over 1500 km. She interpreted this pattern as resulting from selection; supporting information comes from seasonal changes in allele frequencies and laboratory experiments. However, migration studies by Johnston and Heed (1976) indicate that such similarities may result from extensive gene flow over large areas. Resolving the relative roles of selection and migration in determining the genetic structure of *D. nigrospiracula* populations awaits further study.

Thus, three of the four endemic *Drosophila* species have had to shift host plants in order to occupy the major part of the Sonoran Desert. Different host plant preferences within a single species are prime targets for studies in ecological genetics, and differences in allele frequencies of certain enzymes have been correlated with host plant shifts in *D. mojavensis*, *D. arizonensis* and *D. longicornis* (Zouros, 1973; Richardson et al., 1977). However, the shifts are usually associated with geographic effects as well, making the correlations somewhat tenuous until further tests are conducted.

D. pachea does not need to change host plants since senita cactus with plentiful necrotic tissue is found throughout most of the desert. However, in the hot summer months the species exists in extremely low numbers in the southern part of its distribution, even though senita necroses are fairly abundant (Ward, et al., 1975).

Rockwood-Sluss et al., (1973) have studied genetic variation in *D. pachea*. At the ACPH and EST-C loci the genetic variance of allele frequencies within a single locality was as large as the variance among all 11 localities sampled. The localities extend over 800 km. Variance of allele frequencies at the EST-C locus in *D. nigrospiracula* was also as great within a single locality as it was among localities, but this was not true for alleles at the MDH locus (Sluss, 1975). Genetic distance estimates (Rogers, 1972) indicate that the alleles at the MDH locus in *D. nigrospiracula* reflect isolation by distance (Table 2). These data point out the necessity of partitioning dispersal effects from selection effects. *D. pachea* individuals are highly mobile, having 8% to 14% migration rates between cacti (Johnston, 1974).

Ward et al. (1975) have observed a north-south cline in the frequency of a chromosomal inversion in *D. pachea*. The change in frequency closely corresponds to climate and also to differences in the surface-volume ratio of the arms of its host plant. Comparative chromosomal studies currently underway for *D. pachea* in Baja California show that the cline is shifted southward following the major vegetation zones of the peninsula (G. A. Duncan, personal communication). Comparative material from other areas is indispensable for interpreting clinal patterns within a species, and Baja California makes a very good reference point.

Table 2. Rogers' genetic distance estimates and standard errors for EST-C and MDH frequencies between populations of *D. nigrospiracula* and EST-C for *D. pachea*

	D. nigrospiracula		*D. pachea*
Population Level	MDH	EST-C	EST-C
Localized necroses[a]	.033 ± .006	.088 ± .011	.070 ± .012
Seasonal collections[b] at a single locality	.040 ± .006	.065 ± .003	.058 ± .006
All collection[c] localities	.056 ± .003	.087 ± .003	.061 ± .004

[a]Adults reared from 5 samples from a single saguaro *(D. nigrospiracula)*; adults reared from 5 neighboring senita cacti *(D. pachea)*.

[b]Seven samples in 3 years *(D. nigrospiracula)*; 4 samples in 3 years *(D. pachea)*.

[c]Nineteen localities *(D. nigrospiracula)*; 11 localities *(D. pachea)*.

C. Specific Studies with Drosophila mojavensis

I. Host preference and genetics

The shift of *D. mojavensis* from agria in Baja California to organ pipe in Arizona, Sonora, and western Chihuahua has already been mentioned. It is further documented in Table 3, which lists the average number of flies reared per successful collection from each area. These data suggest that agria cactus in Baja California is

Table 3. Mean *D. mojavensis* reared per collection from its two major host plants in the Sonoran Desert. Collections yielding flies are the only ones reported.

	Organ Pipe	Agria	Total
Sonora and Isla Tiburón	225 (20 collections)	255 (9 collections)	234
Baja California	50 (2 collections)	272 (31 collections)	258
Total	210	267	

the most productive habitat and that agria from Isla Tiburón and the region of Desemboque ranks second. The area of greatest heterozygosity for chromosome 2, with four gene arrangements, and chromosome 3, with two gene arrangements, lies along the Gulf Coast region in Baja California (Johnson, 1973; Johnson and Heed, 1977). This chromosomal variation extends across the Gulf of California to Isla Tiburón and the Desemboque region of Sonora. One collection from Desemboque showed the following inversion frequencies for chromosome 2: LP 79%, ST 18%, and BA 4%. A second collection from 25 km north of Punta Chueca contained 98% LP and 2% ST. Both of these areas contain agria cactus (Fig. 3). Three other collections in the Desemboque region where agria was absent proved to be homozygous LP, the karyotype typical for mainland Sonora. All chromosomal heterozygosity of any consequence is associated with agria cactus.

It is assumed that inversions spread throughout a species only by migration but that variants at a single locus can occur and reoccur anywhere by mutation (Powell et al., 1973). These two kinds of genetic markers, if they segregate independently, usually do not exhibit the same pattern of distribution; and the case of *D. mojavensis* is no exception. Zouros (1973) and Zouros and Johnson (1976) demonstrated significant differences in allele frequencies between the mainland (and its nearby islands) and Baja California (and its nearby islands) at four of the seven variable loci measured. The greatest differences were exhibited at the ADH locus located on chromosome 3 of *D. mojavensis*. The allelic frequencies (from data compiled by Zouros (1973) and from the unpublished data of Drs. E. Zouros and E. S. Sluss) are tabulated in Tables 4 and 5 and mapped in Figures 2 and 3.

In Baja California the mean of mean frequencies and their standard deviations are $ADH^F = .971 \pm .026$ and $ADH^S = .029 \pm .026$. In the mainland they are $ADH^F = .151 \pm .065$ and $ADH^S = .849 \pm .065$. The frequencies are extremely stable within the limits of sampling error throughout their respective distributions. This suggests, but does not confirm, uniform but different selection pressures in each area.

The main question here is whether this striking difference in allele frequency results from the host plant shift or the effect of geography. The locality of interest is the Desemboque region where both host plants of *D. mojavensis* are sympatric and supposedly equally inhabitable. Figure 3 illustrates the allele frequencies from collections made from Desemboque to Bahia Kino and also from Isla Tiburón. These collections span a period of three years and include all seasons of the year, yet the frequencies of the two alleles are relatively constant.

Table 4. Allelic frequencies at the ADH locus in *D. mojavensis* from Baja California and nearby islands in the Gulf of California

Map No.	Culture No.	Locality	Date	ADH^S	ADH^F	No. Indep. Genomes	Ecology
1.[d]		Isla San Lorenzo	Spring '71	0	1.00	24	
2.[d]		Isla San Esteban	Spring '71	.06	.94	134	
3.[d]		Isla Tortuga	Spring '71	.09	.91	46	
4.[d]		Isla San Ildefonso	Spring '71	.06	.94	32	
5.[d]		Isla Carmen	Spring '71	0	1.00	20	
6.[d]		Isla Montserrate	Spring '71	.01	.99	208	
7.[d]		Isla San Jose	Spring '71	0	1.00	30	
8.[d]		Los Planes	Spring '70	.05	.95	124	
9.	A420	Punta Prieta	Feb. 17 '73	.02	.98	126	agria[b]
10.	A421	San Ignacio	Feb. 19 '73	.02	.98	176	agria[b]
11.	A422	San Luças	Feb. 21 '73	.01	.98[a]	122	agria[b]
12.	A426	Mulegé (12 klm. N.)	Apr. 15 '73	.04	.96	94	agria[b]
13.	A427	El Coyote	Apr. 16 '73	.04	.96	94	agria[b]
14.	A428	Puerto Escondido	Apr. 16 '73	.04	.95[a]	122	agria[b]
15.	A433	Buena Vista	Apr. 22 '73	.01	.99	274	agria[b]
16.	A587	La Paz (12 klm. W.)	Mar. 9 '76	.05	.95	42	agria[c]
17.	A597	Santiago	Mar. 11 '76	.04	.96	52	agria[c]
18.	A606	Las Flores	Mar. 11 '76	.04	.96	54	agria[c]
19.	A609	San Ignacio	Mar. 15 '76	0	1.00	34	agria[c]
20.	A610	Mulegé	Mar. 16 '76	0	1.00	38	agria[c]

[a] A third allele present

[b] Aspirated from necrotic tissue

[c] Reared from necrotic tissue

[d] From Zouros (1973)

Table 5. Allelic frequencies at the ADH locus in *D. mojavensis* from Arizona, Sonora, and nearby islands in the Gulf of California.

Map No.	Culture No.	Locality	Date	ADH^S	ADH^F	No. Indep. Genomes	Ecology
21.[d]		Organ Pipe Nat. Mon.	Spring '70	.84	.16	130	
22.[d]		Navojoa	Spring '70	.88	.12	96	
23.[d]		Vicam	Spring '70	1.00	0	14	
24.	A410	Desemboque	Jan. 9 '73	.88	.12	170	agria[a]
25.	A411	Red Hills Camp	Jan. 10 '73	.87	.13	88	organ pipe[a]
26.	A412	Punta Chueca (25 klm. N.)	Jan. 11 '73	.88	.12	126	agria[b]
27.	A506	Isla Tiburón	Nov. 11 '73	.84	.16	68	agria[b]
28.	A509	Desemboque	Mar. 18 '74	.84	.16	144	agria[a,c]
29.	A510	Campo Vibora	Mar. 19 '74	.89	.11	152	organ pipe[a,c]
30.	A511	Bahia Kino	Mar. 20 '74	.75	.25	64	traps[c]
31.	A512	Bahía San Carlos	Mar. 22 '74	.86	.14	194	organ pipe[a]
32.	A513	Isla San Pedro	Mar. 20 '74	.71	.29	84	traps[c]
33.	A580	Rancho del Puerto	Jul. 24 '75	.86	.14	216	organ pipe[a,c]
34.	A581	Punta Chueca (25 klm. N.)	Jul. 25 '75	.80	.20	168	agria[a,c]
35.	A700	Bahía San Carlos	May 16 '77	.84	.16	116	organ pipe[a]

[a]Aspirated from necrotic tissue

[b]Reared from necrotic tissue

[c]Banana traps

[d]From Zouros (1973)

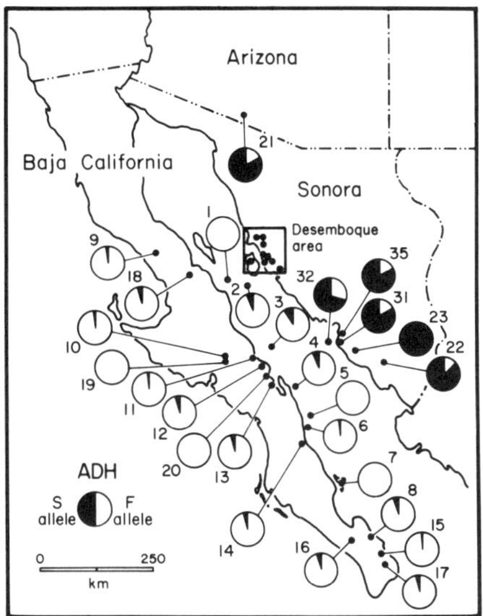

Fig. 2. Allozyme frequencies of the fast (ADH^F) and slow alleles (ADH^S) at the alcohol dehydrogenase locus of *D. mojavensis* in the Sonoran Desert. Numbers refer to localities in Tables 4 and 5. The Desemboque area is enlarged in Figure 3.

The regions of high concentration of agria cactus are marked. Areas both inside and outside of the agria "thickets" contain individual organ pipe cacti. There is no indication of a host plant effect even though five of the nine collection sites were within the agria areas, and two of the samples were taken from adults reared directly from an agria necrosis (Table 5).

At present we have four possible explanations for these observations: (1) there is a host plant effect, but the distribution and abundance of agria in the Desemboque region is not sufficiently large to counteract the effects of immigration from the much larger organ pipe area; (2) there is a geographic effect in the chemistry and/or microorganisms of the two host plants rendering them similar in certain localities; (3) climate is more important than host plants in determining gene frequencies; and (4) epistatic interactions may be maintaining frequency differences on each side of the Gulf in the sense that *D. mojavensis* is presently splitting into two species. These alternatives are not mutually exclusive, e.g., host plant shifts could speed up differentiation. A fifth explanation might claim that the ADH locus is tightly linked to another unidentified locus causing the effect. Even if this were true, the four hypotheses listed above might be operable as well. The swamping effect by migration is entirely possible (Dr. J. S. Johnston, personal communication) even though inversions BA and ST are surrounded by homozygous LP populations in the Desemboque region. Selection may be stronger on chromosome segments than on alleles at a single locus. There is also a little evidence for hypotheses (2) and (4).

II. Host plant alcohols and their relation to survival

The alcohol dehydrogenase locus in *Drosophila* has received detailed attention from a number of investigators, since activity at this

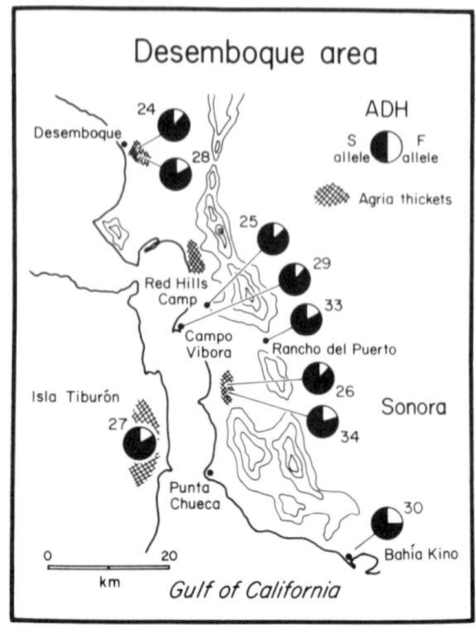

Fig. 3. Allozyme frequencies of the fast (ADH^F) and slow alleles (ADH^S) at the alcohol dehydrogenase locus of *D. mojavensis* in the Desemboque area of the Sonoran Desert. Numbers refer to localities in Table 5. Agria is the host cactus, *Machaerocereus gummosus*.

locus might be related directly to the ecology of the species (see reviews by Clarke, 1975; and Hedrick, et al., 1976). Furthermore, genetic variation at this locus in *D. mojavensis* may be greater than previously recognized since Johnson (1977) found considerable conformational variation when the NAD^+ free form of the protein was examined. In any event, we can now measure the differences and similarities in the alcohol content of necrotic tissue of the various species of host cacti. Table 6 lists

Table 6. Mean and maximum alcohol and acetone concentrations in necrotic host plants of *D. mojavensis* in percent by volume X 10^4. Agria samples from 7 plants from southern Baja California. Organ pipe samples from 6 plants from Organ Pipe National Monument and Sil Nagya, Arizona.

	Agria (12 samples)		Organ Pipe (9 samples)	
	Mean	Maximum	Mean	Maximum
Methanol	88	262	105	212
Ethanol	8	31	142**	361
Acetone	285**	1548	15	39
1-propanol	6	30	71	200
2-propanol	403*	2510	17	56

Significant difference (agria vs. organ pipe) at .05* and .01** by Mann - Whitney U test.

the one to three carbon alcohols and acetone found in the necrotic tissues of agria and organ pipe, determined by gas chromatography by D. C. Vacek (0.2% Carbowax 1500 on 60/80 Carbopack C, Supelco, Inc.). Organ pipe has a significantly higher mean concentration of ethanol and a substantially higher mean concentration of 1-propanol than does agria, while the latter has significantly higher mean concentrations of acetone and 2-propanol. Variances in concentrations are very high; consequently, maximum concentrations are listed instead. No special precautions were taken to preserve the specimens in transit, which may have resulted in some loss of these volatile compounds. Even so, the observed differences correspond well to the following observations.

Starmer et al. (1977) demonstrated a very large increase in longevity of *D. mojavensis* adults when they were held over 1% to 4% solutions of ethanol in water. This result is independent of diet, and the test also demonstrated the retention of mature ovarioles. [Frequencies of ADH^F in Table 2 of Starmer et al. (1977) are higher than those presented in Tables 4 and 5 of this report, apparently because the fast allele is selected for in laboratory cultures.] This increase in survival in the presence of ethanol vapor is at variance with the usual interpretation that (1) an alcoholic environment is deleterious to life, and (2) it is essential for an organism to detoxify alcohols to survive. Others, however, have postulated that flies use ethanol for energy when it is present in the diet (Van Herrewege and David, 1974; Libion-Mannaert et al., 1976).

The discovery that females of strains of *D. mojavensis* from Arizona and Sonora outlived females of strains from Baja California in alcohol tests is of special interest, since it implies that the ADH locus may be an important component of the differential effect. (The order of survival at 2% and 4% ethanol was Sonora females > Sonora males = Baja males > Baja females).

Similar recent tests by E. S. Sluss on homozygous ADH^S and ADH^F genotypes from the same locality (Table 7) show that both genotypes of both sexes significantly outlived the controls in 2% and 4% ethanol vapor, as did both genotypes of females in 2% 2-propanol vapor. Methanol appears to be toxic to the flies. More important, however, in comparisons between the two genotypes, females of ADH^S significantly outlived those of ADH^F in all concentrations of ethanol vapor and in 2% 1-propanol vapor. Statistical comparisons among genotypes of males were not made since the controls were significantly different. High concentrations of ethanol and 1-propanol in necrotic tissues of organ pipe cactus correlate nicely with these findings. Other tests with combinations of primary and secondary alcohols showed that when significant differences in longevity existed between flies from culture A345 from Organ Pipe National Monument, the longer lived individuals were invariably ADH^S. One of this series of tests is shown in Figure 4 in which the first two alcohol tests show a significant difference between females.

Our results partly confirm the results of Ainsley and Kitto (1975) who discovered that *D. melanogaster* larvae homozygous for ADH^S oxidized ethanol, 1-propanol, and ethanol in the presence of 1-propanol, better than larvae homozygous for ADH^F and mostly better than ADH^{FS} heterozygotes. On the other hand, ADH^{FF} larvae oxidized 2-propanol and ethanol in the presence of 2-butanol better than ADH^{SS} and ADH^{FS} larvae. [Figure 5 in Ainsley and Kitto (1975) is mislabeled; the alcohols 1-propanol and 2-propanol should be reversed (Dr. R. Ainsley, personal communication)].

Table 7. Mean days survival, and their standard deviations, of ADH homozygous genotypes of four-day old *D. mojavensis* in atmospheric alcohol without food. Flies from Organ Pipe National Monument, Arizona (A345).

	Females				Males			
	ADH^S	N	ADH^F	N	ADH^S	N	ADH^F	N
Water	11.2 ± 5.4	50	11.5 ± 2.9	49	10.8 ± 2.0	60	9.1 ± 1.6	50
2% Ethanol	15.3 ± 5.9**	28	12.9 ± 5.9*	49	13.5 ± 5.3**	40	12.4 ± 1.7**	40
4% Ethanol	18.0 ± 2.0**	10	15.4 ± 4.2**	20	14.1 ± 5.2**	10	11.9 ± 1.9**	10
8% Ethanol	11.8 ± 6.0	21	10.3 ± 2.1*	20	9.6 ± 3.1	20	9.3 ± 0.5	10
2% 1-Propanol	14.1 ± 10.4*	20	11.1 ± 2.9	20	11.2 ± 3.4	20	10.4 ± 4.0*	19
2% 2-Propanol	12.8 ± 2.0**	27	13.0 ± 1.5**	30	12.6 ± 8.1	17	8.5 ± 2.7	20
2% Methanol	10.8 ± 2.1	5	9.3 ± 1.2**	10	8.3 ± 2.2**	9	8.4 ± 0.5*	10

Significant at .01* and .001** each column vs. water control.

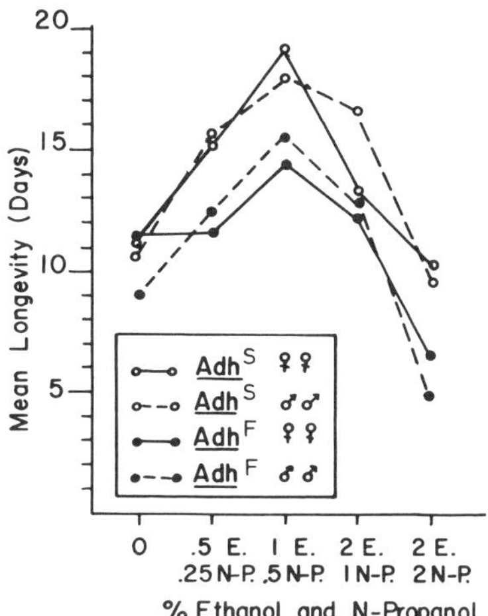

Fig. 4. Relative longevity of genotypes and sexes of *D. mojavensis* held without food over increasing strengths of mixtures of ethanol (E) and 1-propanol (N-P). ADH^S and ADH^F are homozygous individuals for the slow and fast alleles at the alcohol dehydrogenase locus. Culture A345 from Organ Pipe National Monument, Sept. 1971.

In summary, the ADH^F genotype seems to work on a secondary alcohol while the ADH^S genotype evidently prefers two primary alcohols. These results are in accord with the allele frequencies observed in *D. mojavensis* and with the alcoholic composition of the two substrates. They do not precisely coincide with the alcohol vapor tests shown in Table 7 since *D. mojavensis* females homozygous for ADH^F did not outlive ADH^S females in 2% 2-propanol. However, this was the only useable alcohol in which the two genotypes had equal survival. Future tests utilizing cultures derived from Baja California instead of the mainland may show distinct differences in favor of ADH^F with 2-propanol. Starmer et al. (1977) lists other environmental factors that could favor the fast allele in Baja California. Furthermore, the ADH^{FS} heterozygote of this species has yet to be tested in an *in vivo* experiment.

Additional gas-chromatographic analyses performed by D. C. Vacek on the necrotic tissues from agria collected at Punta Chueca, Sonora, and organ pipe from San Carlos Bay, Sonora, revealed interesting new patterns. Several organ pipe profiles were more like agria, showing higher peaks for 2-propanol and acetone, while several agria profiles mimicked organ pipe by having higher peaks for methanol and ethanol or 1-propanol. Other samples were intermediate, showing high peaks for many of the alcohols and ketones. This kind of variation may be the determining factor maintaining both alleles at the ADH locus in populations on both sides of the Gulf, and it illustrates the possibility that hypothesis (2), discussed earlier, may be valid. Since the proportion of necroses sampled which contained adults and/or larvae of *D. mojavensis* is less than 50%, we do not know which patterns are attractive and which ones are not. This is an area for future investigation.

III. Survival in other species

Data from other species of *Drosophila* breeding in other columnar cacti provide an interesting contrast to the story outlined above. Table 8 shows that concentrations of useable alcohols are low in

Table 8. Mean and maximum alcohol and acetone concentrations in necrotic host plants of *D. pachea* (senita), from southern Baja California, and *D. nigrospiracula* (saguaro), from southern Arizona, in percent by volume X 10^4.

	Senita (4 samples)		Saguaro (4 samples)	
	Mean	Maximum	Mean	Maximum
Methanol	205	237	202	304
Ethanol	13	27	15	41
Acetone	32	36	11	31
1-propanol	25	54	30	93
2-propanol	17*	32	0	1

Significant at .05* senita vs. saguaro by Mann - Whitney U test.

sample necroses from senita and saguaro. The four samples of senita are from four different plants, while the four samples of saguaro are all from one plant. Comparison of the total mean percentages of useable alcohols (ethanol, 1-propanol and 2-propanol) from all cacti studied is: agria .0417%, organ pipe .0230%, senita .0055%, and saguaro .0045%. Methanol is relatively more prevalent in senita and saguaro than in organ pipe and agria, and this alcohol cannot be utilized by any desert *Drosophila*. We have shown that adults of *D. nigrospiracula*, which breed in saguaro, cannot utilize ethanol to increase longevity (Table 9); in fact, the 4% and 8% concentrations are toxic.

Table 9. Mean days survival (± S. D.) of eight-day old *D. nigrospiracula* in atmospheric ethanol without food. Ten flies and two replicas each test.

	Females	Males
Water	6.35 ± 1.39	6.06 ± 1.84
2%	6.40 ± 0.88	6.10 ± 1.36
4%	3.05 ± 2.16*	2.70 ± 2.22*
8%	< 1.00*	< 1.00

Significant at .01* each column vs. water control

The flies were eight days old when tested and had never been subjected to anesthesia of any kind. The techniques were the same as those used by Starmer et al. (1977).

Preliminary tests at room temperature with adults of *D. pachea*, which breed in senita, showed that the mean longevity of flies collected at Desemboque, Sonora, increased slightly from 12.9 days (over water) to 14.4 days (over 2% ethanol). However, there was no increase in longevity of adults of the same species descended from flies collected at Zaragosa, Sinaloa. This test showed 10.3 days over water vs. 10.7 days over 2% ethanol.

Thus, there appears to be a reasonable correlation between the presence or absence of one-to-three carbon alcohols in rot pockets of the host plants and the ability of each resident species of *Drosophila* to utilize them. The situation parallels the finding of McKenzie and Parsons (1972), who demonstrated an ecological separation between *D. melanogaster* and *D. simulans* on the basis of their ability to tolerate alcohol. McDonald and Avise (1975) demonstrated a wide range of activity of the ADH enzyme as well as its ontogenetic pattern among nine species of *Drosophila* in the subgenus *Sophophora*. These authors also found a correlation between the level of adult ADH activity and the ability to survive higher concentrations of 2-propanol, and they suggested that at least one of the major functions of the ADH enzyme is for direct adaptation to alcohol in the environment.

The differences in the types and amount of alcohols among the four species of cacti are due in part to the differences in their yeast flora. A total of 14 species of yeasts have been isolated from 100 necrotic samples of agria, and 12 species have been isolated from 30 necrotic samples of organ pipe (Starmer et al., 1976; Heed et al., 1976; and Starmer, personal communication). Only two of these yeast species, *Torulopsis sonorensis* and *Candida tenuis*, ferment glucose. Table 10 shows the comparative distribution of the two yeasts among the four cacti that have been analyzed for their alcohol contents; cacti with higher alcohol contents usually contain the fermentative yeasts. The other microorganism known to ferment glucose is *Erwinia*

Table 10. Estimates of minimal frequency of two fermentative yeast species among four species of cacti.

	Agria	Organ Pipe	Senita	Saguaro
Number of plants sampled	100	30	62	16
Torulopsis sonorensis	.40	.53	.05	.06
Candida tenuis	.03	.23	0	0

carnegieana, the soft-rot bacterium which probably initiates the majority of necroses. Cultures of this organism show noticeable peaks of ethanol and acetone when grown in a fermentation medium of yeast extract and 2% glucose (D. C. Vacek, personal communication).

IV. Alcohol metabolism

The *D. mojavensis* story established to date suggests that the frequency of the two main alleles at the ADH locus are responding to direct environmental effects, and one of its most intriguing aspects concerns the possible use of gaseous ethanol as a source of energy. Circumstantial evidence suggests that ADH is the enzyme primarily responsible for this ability. Therefore we began work with trace-labeled ethanol to elucidate the role of this alcohol in metabolism.

The appropriate experiments were conducted by E. S. Sluss. She demonstrated that $^{14}CO_2$ is produced when *D. mojavensis* adults are held over 1 C^{14} labeled ethanol. Females generated more label than males [2135 vs. 1708 disintegrations per minute (dpm)] after a six day test. Axenic flies do not differ from non-axenic ones in the amount of $^{14}CO_2$ generated. The details of this and other tests will be published at a later date, but the main results show that (1) *D. mojavensis* metabolizes ethanol obtained from the atmosphere, and (2) females metabolize greater amounts than males. Both results are concordant with the findings discussed above.

As a further indication that the ADH locus is involved in metabolism, null mutants of two strains of *D. melanogaster* were compared to wild type in the labeled ethanol test. Even though nulls began to die shortly before analysis, one strain (axenic b ADH^{n-4} of Sofer) produced 858.0 dpm/fly of $^{14}CO_2$ compared to wild type which produced 1532.6 dpm/fly after three days. Similar results were obtained with the second strain (ADH^{n-1} of Grell). These results indicate that other systems in addition to ADH may function to metabolize ethanol.

V. Sexual isolation.

As stated earlier, one hypothesis which might account for the differences in allele frequencies observed on both sides of the Gulf may be that *D. mojavensis* is differentiating into two species. Some evidence for this has been provided by Zouros and d'Entremont (1974), who demonstrated a significant preference for homogamic matings when females had a choice between Baja males and Sonora males. However, the case is far from simple. In an exhaustive analysis of sexual isolation between and within strains of *D. mojavensis* and *D. arizonensis*, Wasserman and Koepfer (1977) showed that only mainland *D. mojavensis* were responsible for the observed isolation. They further demonstrated a small but significant level of isolation between four of six intraspecific matings of *D. mojavensis*. Three of the four matings were between flies from localities in Baja California and Sonora. It appears that mainland *D. mojavensis* are capable of responding to the apparent pressure from a closely related sympatric species. Peninsular *D. mojavensis* provide a nice control since *D. arizonensis* is not present, at least in the stressful area of the desert. These observations confirm that different selection pressures other than the ones associated with host plant choice and climate exist on both sides of the Gulf. Whether sexual selection has modified the gene frequencies discussed here remains to be determined.

Acknowledgments. This paper is dedicated in the memory of Sarah Bedichek Pipkin who enlightened *Drosophila* workers everywhere on the systematics and ecological genetics of her favorite organism.

I thank the following persons for their kind permission to use their unpublished data and for their interest in the project: Garry A. Duncan, Allen C. Gathman, Arthur C. Gibson, Karl E. Horak, Margaret C. Jefferson, William R. Johnson, J. Spencer Johnston, Henry W. Kircher Robert L. Mangan, Richard H. Richardson, E. Susan Sluss, William T. Starmer, Don C. Vacek, Marvin Wasserman, Richard Williams and E. Zouros. I am indebted to Jean S. Russell and Alexander Russell Jr., for their unfailing aid in the laboratory and the field. W. H. Sofer and R. F. Grell kindly supplied the null mutants of *D. melanogaster*, and A. C. Gibson, H. W. Kircher and W. T. Starmer made valuable corrections to the manuscript. This work is supported by the National Science Foundation.

References

Ainsley, R., Kitto, G.F.: Selection mechanisms maintaining alcohol dehydrogenase polymorphisms in *Drosophila melanogaster*. In Isozymes II. Physiological Function. (ed. C. Markert). New York: Academic Press, 1975, pp. 733-743.

Clarke, B.: The contribution of ecological genetics to evolutionary theory: detecting the direct effects of natural selection on particular polymorphic loci. Genetics 79:101-113. (1975).

Cooper, J.W.: Genetic and cytological studies of *Drosophila nigrospiracula* in the Sonoran Desert. M.S. Thesis, University of Arizona, Tucson. (1964).

Fellows, D.P., Heed, W.B.: Factors affecting host plant selection in desert-adapted cactiphilic *Drosophila*. Ecology 53:850-858. (1972).

Gibson, A.C., Horak, K.E.: Systematic anatomy and phylogeny of Mexican columnar cacti. Ann. Missouri Bot. Gard. (in press, 1977).

Hedrick, P.W., Ginevan, M.E., Ewing, E.P.: Genetic polymorphism in heterogeneous environments. Ann. Rev. Ecol. Syst. 7:1-32. (1976).

Heed, W.B.: A new cactus-feeding but soil-breeding species of *Drosophila* (Diptera: Drosophilidae). Proc. Ent. Soc. Wash. 79:649-654. (1977).

Heed, W.B., Starmer, W.T., Miranda, M., Miller, M.W., Phaff, H.J.: An analysis of the yeast flora associated with cactiphilic *Drosophila* and their host plants in the Sonoran Desert and its relation to temperate and tropical associations. Ecology 57: 151-160. (1976).

Johnson, G.B.: Stabilization by cofactor binding of structural flexibility among ADH isozyme variants in *Drosophila*. Genetics 86:33 (Suppl.). (1977).

Johnson, W.R.: Chromosome variation in natural populations of *Drosophila mojavensis*. M.S. Thesis, University of Arizona, Tucson. (1973).

Johnson, W.R., Heed, W.B.: Chromosomal polymorphism in the desert-adapted species, *Drosophila mojavensis*. Evolution (in press, 1977).

Johnston, J.S.: Dispersal in natural populations of the cactiphilic *Drosophila pachea* and *D. mojavensis*. Genetics 77:32-33 (Suppl.). (1974).

Johnston, J.S., Heed, W.B.: Dispersal of desert-adapted *Drosophila*: The saguaro-breeding *D. nigrospiracula*. Amer. Natur. 110:629-651. (1976).

Kaneshiro, K.Y., Carson, H.L., Clayton, F.E., Heed, W.B.: Niche separation in a pair of homosequential *Drosophila* species from the island of Hawaii. Amer. Natur. 107:766-774. (1973).

Libion-Mannaert, M., Delcour, J., Deltombe-Lietaert, M.C., Lenelle-Montfort, N., Elens, A.: Ethanol as a "food" for *Drosophila melanogaster*: Influence of the *ebony* gene. Experientia 32: 22-24. (1976).

McDonald, J.F., Avise, J.C.: Evidence for the adaptive significance of enzyme activity levels: Interspecific variation in α-GPDH

and ADH in *Drosophila*. Biochem. Genet. 14:347-355. (1975).
McKenzie, J.A., Parsons, P.A.: Alcohol tolerance: An ecological parameter in the relative success of *Drosophila melanogaster* and *Drosophila simulans*. Oecologia 10:373-388. (1972).
Patterson, J.R., Stone, W.S.: Evolution in the Genus *Drosophila*. New York: The Macmillan Co., 1952, 610 pp.
Powell, J.R., Levene, H., Dobzhansky, T.: Chromosomal polymorphism in *Drosophila pseudoobscura* used for diagnosis of geographic origin. Evolution 26:553-559. (1973).
Richardson, R.H., Smouse, P.E., Richardson, M.R.: Patterns of molecular variation. II. Associations of electrophoretic mobility and larval substrate within species of the *Drosophila mulleri* complex. Genetics 85:141-154. (1977).
Rockwood-Sluss, E.S., Johnston, J.S., Heed, W.B.: Allozyme genotype-environment relationships. I. Variation in natural populations of *Drosophila pachea*. Genetics 73:135-146. (1973).
Rogers, J.S.: Measures of genetic similarity and genetic distance. Univ. of Texas Publ. 7213:145-153. (1972).
Shreve, F., Wiggins, I.L.: Vegetation of the Sonoran Desert. Vol. 1. Stanford, California: Stanford University Press, 1964.
Sluss, E.S.: Enzyme variability in natural populations of two species of cactophilic *Drosophila*. Ph.D. Dissertation, University of Arizona, Tucson. (1975).
Starmer, W.T., Heed, W.B., Miranda, M., Miller, M.W., Phaff, H.J.: The ecology of yeast flora associated with cactiphilic *Drosophila* and their host plants in the Sonoran Desert. Microbial Ecology 3:11-30. (1976).
Starmer, W.T., Heed, W.B., Rockwood-Sluss, E.S.: Extension of longevity in *Drosophila mojavensis* by environmental ethanol: differences between subraces. Proc. Natl. Acad. Sci. 74:387-391. (1977).
Van Herrewege, J., David. J.: Utilisation de l'alcohol éthylique dans le métabolisme énergétique d'un insects: Influence sur la durée de survie des adultes de *Drosophila melanogaster*. C.R. Acad. Sci. Ser. D 279:335-338. (1974).
Ward, B.L., Starmer, W.T., Russell, J.S., Heed, W.B.: The correlation of climate and host plant morphology with a geographic gradient of an inversion polymorphism in *Drosophila pachea*. Evolution 28:565-575. (1975).
Wasserman, M., Koepfer, H.R.: Character displacement for sexual isolation between *Drosophila mojavensis* and *Drosophila arizonensis*. Evolution 31:812-823. (1977).
Wiggins, I.L.: The origin and relationship of the land flora. Syst. Zool. 9:148-165. (1960).
Zouros, E.: Genic differentiation associated with the early stages of speciation in the *mulleri* subgroup of *Drosophila*. Evolution 27:601-621. (1973).
Zouros, E., D'Entremont, D.: Sexual isolation among populations of *D. mojavensis* race B. Dros. Inf. Serv. 51:112-113. (1974).
Zouros, E., Johnson, W.: Linkage disequilibrium between functionally related enzyme loci of *Drosophila mojavensis*. Can. J. Genet. Cytol. 18:245-254. (1976).

Microspatial Genetic Differentiation in Natural Populations of *Drosophila*

R.C. RICHMOND

A. Introduction

Geographic variation in morphological characters and gene frequencies within species has long been studied by ecological and evolutionary geneticists. Indeed, the generalizations resulting from investigations of intraspecific variation have been canonized as "ecogeographical rules" (Mayr, 1963). The study of geographic variation led to the formulation of the allopatric or geographic model of speciation which is accepted by most evolutionary biologists as the predominant means of speciation among sexually reproducing species. This emphasis on how species are adapted to general climatic and geographical variation has often diverted attention from the problems of adaptation to local environmental heterogeneity.

The way in which a population is adapted to environmental heterogeneity depends upon the "grain" of the environment, the relative frequency of environmental "patches", and the fitnesses of the various genotypes present in the population in each patch (Levins, 1968). If the environment is coarse-grained so that an individual may spend its whole reproductive life in one patch, selection may act in a disruptive manner favoring one genotype in one patch, another in a second patch and so on. Depending upon the degree of migration between patches and subsequent gene flow, the population may become subdivided and exhibit spatial differentiation of gene frequencies. This argument makes intuitive sense and provides a mechanism by which a population may become genetically differentiated other than by geographic isolation. Evidence that microspatial heterogeneity in gene frequencies exists in natural populations is beginning to accumulate. A number of both selective and non-selective processes may give rise to spatial differentiation in gene frequencies, and some of these are considered here in conjunction with the analysis of spatial heterogeneities in gene frequencies within *Drosophila* populations.

Our objective here is to review the evidence that *Drosophila* are capable of perceiving and responding to environmental heterogencities, and that this heterogeneity can affect genetic variation. We will focus on a particular example of a *Drosophila* population which does show microspatial differentiation in gene frequencies and consider the possible explanations for this result. Population structuring as revealed by microspatial differentiation in gene frequencies may be of substantial evolutionary importance.

B. The Significance of Population Structure

If individual members of a population perceive their environment as consisting of a series of distinguishable habitats, (i.e., fine-grained) it is clearly to the population's and species' advantage to utilize all aspects of the environment rather than specializing on one or a few of the available niches. Levene (1953) and later Dempster (1955) and Li (1955) formalized this idea by showing that allelic variants at a locus could be maintained in a population in the absence of overdominance if different alleles were favored in different niches. Levene drew attention to the possibility that the existence of multiple niches within the environment occupied by a population might lead to mating within niches and thus to a spatially structured population.

If many natural populations are not homogeneous genetic units but rather consist of partially isolated subpopulations, such structuring may allow for selection between subpopulations. This process might permit the composite population to move from one adaptive peak to another without the necessity of transversing the adaptive valley between. Without population structuring, such evolutionary shifts would not occur in large, panmictic populations (Crow and Kimura, 1970). While this concept has received theoretical attention, the prevailing view among evolutionary biologists has been that populations and species are cohesive units largely maintained by gene flow (Dobzhansky, 1955; Mayr, 1963). Ehrlich and Raven (1969) have questioned this assumption and present a variety of data to support the hypothesis that gene flow both within and between populations may be relatively unimportant in maintaining the genetic structure of populations and species. Thus the spatial structuring of populations may have a profound influence on the maintenance of genetic variability within the composite population and its potential for large adaptive shifts. If population structuring were shown to be a common phenomenon in nature, the established notion that the unit of evolutionary significance is a Mendelian population would be displaced by the hypothesis that the local population or deme is of primary importance.

C. Microspatial Variation in Gene Frequencies in *Drosophila* Populations

A growing series of both field and laboratory studies provide evidence that *Drosophila* are capable of discriminating among a variety of environmental variables, and that these variables can affect the degree of genetic variation maintained in artifical populations. Field investigations of *Drosophila* populations reveal that microspatial variation in gene frequencies is often present.

I. Studies of habitat preference

A variety of laboratory investigations have shown that *Drosophila* are capable of discriminating among the various physical and biological attributes of an artifical environment. Waddington et al. (1954) demonstrated some time ago that mutant lines of *D. melanogaster* showed clear preferences for areas within an experimental apparatus in which temperature, humidity, and light were varied. Responses to two of these variables have since been shown to be under polygenic control (Richmond, unpublished results; Rockwell and Seiger, 1973). Richardson and Johnston (1975a) have demonstrated that *D. mimica*, an Hawaiian species, shows quite precise responses to wind velocity in the laboratory and that these behaviors help to explain its microdistribution under natural conditions. Food preferences among various species of *Drosophila* have been repeatedly demonstrated (Heed, 1971).

The precision with which *Drosophila* females select sites for oviposition is remarkable. Kaneshiro et al. (1973) have shown that oviposition behavior is an essential factor maintaining niche separation in a pair of sympatric species of *Drosophila* from the island of Hawaii. Both species utilize the same sap flows from a single species of tree as feeding sites. However, oviposition and larval development sites are completely separate in the two species. Females of *D. heedi* oviposit in leaf litter or soil moistened by dripping sap, while females of *D. silvarentis* utilize sap flow areas on the trunk and branches of the tree. Carson (1974) has described an "evolutionary innovation" in three species of *Drosophila* which have independently evolved the ability to use the nephric duct or gill chambers of tropical land crabs as oviposition and larval feeding sites!

Richardson and Johnston (1975 a,b) have demonstrated that habitat selection is an important behavioral mechanism maintaining microspatial isolation between species and sexes of *Drosophila* in a 60m x 70m area within a "kipuka" on the island of Hawaii. The species were shown to differ in their substrate preferences and light tolerance. Females of *D. mimica* are less light tolerant than males and are restricted to shaded areas. Similar examples of both lateral and vertical differences in the spatial distributions of the sexes were found for two other species studied. These and many other studies have demonstrated that *Drosophila* are capable of perceiving and responding to microclimatological differences within their immediate environment. Microspatial separation of species or sexes is often the result, and such separation is likely to result in a mosaic of selection patterns. We next investigate whether such micro-environmental heterogeneity can influence genetic variation within a *Drosophila* population.

II. The Effect of Environmental Heterogeneity on Genetic Variation

Laboratory experiments designed to investigate the adaptive significance of allozyme polymorphisms in *Drosophila* bear directly on the question of whether microspatial variation in a population's environment is capable of influencing genetic variation. Three separate sets of experiments have been reported which provide the needed data (Powell, 1971; McDonald and Ayala, 1974; Powell and Wistrand, cited in Hedrick et al., 1976). All these experiments report very similar results. Population cages of *Drosophila* were maintained under constant conditions, or conditions in which the population was exposed to 1, 2, 3 or 4 variable environmental factors. Average allozyme-gene heterozygosites in the cages were analysed after a period of several generations. The average proportion of loci heterozygous per individual showed a direct relationship to the number of variable environmental factors. Beardmore (1970) and his coworkers have obtained similar results for a single allozyme locus in *D. melanogaster*. These data demonstrate that under laboratory conditions, small scale environmental variation does have a direct effect on genetic variability measured by allozyme-gene heterozygosities.

Both laboratory and field investigations show that *Drosophila* respond to microclimatological variation in their environment and that such variation can act as a selective force. We now ask whether there is evidence for microspatial differentiation of gene frequencies in natural populations of *Drosophila*.

III. Field Studies of Microspatial Variation in Gene Frequencies

Wallace (1966) investigated the frequency of allelism of lethals in samples of *D. melanogaster* collected from four baits separated from each other by a distance of 30 meters. Although Chi-square tests for homogeneity in the frequencies of allelism between collection sites were not significant, there was a highly significant relationship between the allelism rate of alleles identical by descent and the square root of the distance between baits. This result strongly suggests that populations which are separated by as little as 30 meters can be genetically differentiated. Wallace's data (1970) on the extent of dispersal in *D. melanogaster* support this conclusion.

Richardson (1968, 1969) investigated local differences in the frequencies of alleles at an esterase locus in *D. aldrichi*. An area 3 km x 2.5 km was shown to contain two genetically distinct populations as judged by the frequencies of esterase alleles. Migration rates estimated by the use of a self-marking technique indicate that *D. aldrichi* move at most only a few meters a day.

In Australia, McKenzie and Parsons (1972) have shown that the sibling species, *D. melanogaster* and *simulans*, are differently distributed with respect to the presence of alcohol in the environment. *D. melanogaster* is more resistant to alcohol and has been collected in wine cellars, whereas both *D. simulans* and *melanogaster* can be collected immediately outside the cellars. Since *D. melanogaster* occurs in both the alcohol rich environment of the cellars and the relatively alcohol deficient environment outside the cellars, an opportunity for genetic differentiation exists if selection pressures are strong enough and gene flow is restricted. McKenzie and Parsons (1974) have shown that strains of *D. melanogaster* collected inside a wine cellar are significantly more resistant to alcohol than strains collected outside the cellar. This genetic difference is not associated with the alcohol dehydrogenase polymorphism in this species, but is apparently a result of the interactions of several loci involving both additive and dominance effects. Genetic differentiation appears to have occurred in this population in spite of gene flow, although the evidence for gene flow is only circumstantial.

A series of recent investigations have shown that the frequencies of chromosome inversions in several species of *Drosophila* can be spatially differentiated. The variances of inversion frequencies from collections of *D. subobscura* were shown by Krimbas and Alevizos (1973) to be homogeneous for collections taken from the same traps at different times but heterogeneous for collections taken from different traps at the same time. In some cases the traps were separated by as little as 50 m. An analysis of seven allozyme loci did not provide evidence for microdifferentiation, possibly indicating that the differences found for inversions were due to a mosaic pattern of selection.

Stalker (1976) studied the frequencies of inversions in *D. melanogaster* from flies collected from rotting oranges and grapefruit in closely adjacent areas. His results show that populations which are separated by as little as 18 m may be genetically different with respect to inversions in all four of the major autosomal chromosome arms.

The frequencies of both 3rd chromosome gene arrangments and allozyme genes in *D. persimilis* were studied in an area about 700 x 800 m by Taylor and Powell (1977). Inversion frequencies were found to differ among ecologically diverse areas such as a meadow, open woods and moist woods. The frequencies of electromorphs at polymorphic allozyme loci were also found to be heterogeneous among the areas studied. Studies of dispersal in these species indicate that individual flies are capable of transversing the distances between the areas studied in a period of a few days. Whether or not this dispersal represents effective gene flow is unclear.

Relatively few species of *Drosophila* have been studied in which the breeding substrate has been clearly identified. Among the exceptions to this statement are several species of the Sonoran Desert region studied extensively by Heed and his colleagues. Careful studies of dispersal in one of these species, *D. nigrospiracula*, shows that it is capable of moving as far as 1000 m in a single day to locate a new food source (Johnston and Heed, 1976). Rockwood-Sluss et al. (1973) studied microspatial differentiation in *D. pachea*, a species whose ecology is similar to that of *D. nigrospiracula*, by analysing allozyme gene frequencies in flies collected within walking distance in a single stand of senita cactus. The variance in gene frequencies at two loci among collections from a single locality was equal to that among populations separated by 500 miles. It is likely that the extreme dispersal rates characteristic of *D. nigrospiracula* also hold for *D. pachea* and explain the homogeneity in gene frequencies. The available data on microspatial differentiation in *Drosophila* reviewed above and discussed in some detail below for *D. affinis* are summarized in Table 1.

IV. Spatial and Temporal Variation in Allozyme-Gene Frequencies in *Drosophila affinis*

During the past several years, our laboratory has been engaged in a study of microdifferentiation in *Drosophila affinis*. *D. affinis* is a member of the obscura group of *Drosophila* and is widely distributed in North America (Miller, 1958). There is no evidence for any incipient reproductive isolation within this species, and an analysis of allozyme variation in geographic samples has revealed the now familiar pattern of similarity in gene frequencies in different populations (Richmond, unpublished data). We have investigated the pattern of spatial and temporal variation in allozyme-gene frequencies in a single population.

1. Methods

Collections of *D. affinis* were made from a site near Unionville, Indiana. The locality is a second-growth forest dissected by ravines containing seasonal streams. A 40m x 75m grid, marked at five meter intervals, was located on the side of a ravine. It encompasses one side of the stream bed and approximately one-third of the slope of the ravine. The lower third of the grid is an approximately level area next to the watercourse of the stream. The central area is a steeply sloping region containing relatively little understory vegetation. The top third of the grid is a more gently sloping area covered by numerous trees and abundant small vegetation. Small fermenting banana baits nine cm in diameter were placed at five

Table 1. Studies of Microspatial, Genetic Differentiation in *Drosophila*

Species	Genetic System Studied	Minimum Distance Between Differentiated Populations	References
affinis	6 allozyme loci	25 meters	Richmond (1977)
aldrichi	1 allozyme locus	400 meters	Richardson (1968)
melanogaster	allelism of lethals	30 meters	Wallace (1966)
melanogaster	alcohol resistance	≃ 10 meters	McKenzie and Parsons (1977)
melanogaster	chromosome inversions	18 meters	Stalker (1976)
pachea	2 allozyme loci	No differentiation	Rockwood-Sluss et al. (1973)
persimilis	chromosome inversions 9 allozyme loci	≃200 meters ≃200 meters	Taylor and Powell (1977)
subobscura	chromosome inversions 7 allozyme loci	50 meters No microdifferentiation	Krimbas and Alevizos (1973)

meter intervals throughout the grid for a total of 120 traps. Collections were made during the summer months of 1973 and 1974 and returned to our laboratory for the analysis of allozyme variation at six polymorphic loci.

2. Results

The use of many small baits and a portable vacuum cleaner which allowed rapid collection of flies is assumed to have retained any local differentiation which may have been present among the *Drosophila* captured. This assumption is supported by three observations. (1) An average of only 5.3 Drosophilids were collected per trap. (2) Heterogeneities in the distribution of *D. affinis* and the related species, *athabasca* and *algonquin* were detected between the upper and lower areas of the grid as summarized in Table 2. (3) Heterogeneities in gene frequencies for *D. affinis* were observed at some loci. The small numbers of flies collected at each bait has necessitated the grouping of data by areas within the grid. We chose to group the collections within the three natural subdivisions of the grid described above.

Table 2. Numbers of male individuals of three *Drosophila* species collected in the upper and lower areas of the sample grid

Species	Upper Area	Lower Area	Total	Chi-square	P
affinis	628	870	1552	22.8	< .005
athabasca	219	170	389	6.2	< .025
algonquin	16	36	52	7.7	< .010

Of the six polymorphic loci studied in *D. affinis*, the leucine aminopeptidase and alkaline phosphatase loci showed neither temporal nor spatial heterogeneity. Four loci, Est-6, Est-6A, Est-7 and octanol dehydrogenase (ODH), provided evidence for spatial and/or temporal heterogeneity in allozyme gene frequencies. The Est-6 and Est-6A loci proved to be spatially and temporally heterogeneous.

Allele frequencies at the Est-6 locus are provided in Table 3. Likelihood ratio tests, G values, are given in the table to test for the homogeneity of allele frequencies over the various sample dates within areas of the collection grid. The several collections from the upper and middle areas of the grid are temporally homogeneous, but those from the lower area are not. A combined test of temporal heterogeneity obtained by summing the G values and degrees of freedom for all the Est-6 data is significant. This temporal heterogeneity in allele frequencies prevents an overall comparison of allele frequencies from the three areas of the grid. Accordingly a series of five tests (one for each of the sample dates) have been made to compare allele frequencies in the three areas of the grid. Of these five comparisons, the July 1973 samples are statistically heterogeneous (G=15.88, 6df, P=0.014). The predominant allele at the Est-6 locus, Est-6^{102}, shows a fairly consistent pattern of increase in frequency between the lower and upper areas of the grid (Table 3). This apparent association between the position of the collection and the

Table 3. Electromorph frequency distributions at the Est-6 locus.

Location	Date	No. Flies	Electromorphs 100	101	102	103	Rare	G	P[d]
Lower[a]	7/73	69	.058	.174	.449	.290	.029	29.64	0.003
	8/73	36	.028	.250	.583	.083	.056		
	4/74	49	.102	.265	.531	.102	0		
	5/74	50	.100	.700	.700	.100	0		
	6/74	49	.184	.204	.551	.061	0		
Middle[b]	7/73	77	.026	.143	.597	.195	.039	2.58	> 0.25
	8/73	13	0	.154	.769	0	.077		
Upper[c]	7/73	101	0	.139	.703	.099	.059	14.78	> 0.25
	8/73	21	.048	.048	.762	.143	0		
	4/74	50	.040	.200	.640	.100	.020		
	5/74	48	.063	.188	.687	.062	0		
	6/74	49	.102	.163	.633	.020	.082		

[a]Electromorphs 103 and Rare combined for G test

[b]Electromorphs 100, 103 and Rare combined for G test

[c]Electromorphs 100 and Rare combined for G test

[d]Combined probability of temporal heterogeneity G=47.00, 26DF, $P < 0.01$

frequency of this esterase allele is significant as tested by a Spearman's rank correlation coefficient ($r_s = 0.61$, $t = 2.41$, 10df, $0.05 > p > 0.02$).

An analysis of temporal variation at the Est-6A locus revealed substantial heterogeneity within each of the areas of the grid. Therefore, comparisons of spatial heterogeneity in allele frequencies have been restricted to collections made on the same date. These tests are summarized in Table 4 and reveal that two of four samples showed spatial heterogeneity. A composite test obtained by summing the G values and degrees of freedom is significant.

These results clearly show that collections of *Drosophila* within a 40 x 75 m. area can be genetically differentiated as determined by heterogeneities in allozyme-gene frequencies. Other data from *Drosophila* reviewed above demonstrate that microgeographic heterogeneities in gene frequencies have been detected in five of the six *Drosophila* species examined to date. Our next task is to attempt to distinguish among the possible hypotheses which can account for these results.

V. Analysis of microspatial variation in gene frequencies

At least four factors have been identified which can result in microspatial heterogeneity in gene frequencies (Cavalli-Sforza and Bodmer, 1971; Taylor and Powell, 1977). They are genetic drift, natural selection which varies as a function of the location of an individual within the environment, non-uniform dispersal of different

Table 4. Likelihood ratio tests, G, for spatial heterogeneity in allele frequencies at the Est-6A locus

Sample Date	Number of areas compared	DF	G	P
7/73	3	14	43.8	< .005
8/73	3	8	13.5	.095
5/74	2	6	19.7	< .005
6/74	2	5	8.7	< .10
Total		33	85.7	< .005

genotypes, and habitat selection which is genetically conditioned. We will consider below the possibility that each of these factors or some combination of factors can account for the microgeographic differences in gene frequency which are observed in *Drosophila* populations.

1. Genetic Drift

If natural populations of *Drosophila* are composed of a series of small, partially isolated demes, then genetic drift within each deme could give rise to spatial differentiation of gene frequencies. The action of genetic drift is often difficult to establish because its efficacy depends upon the magnitude of two demographic parameters, population sizes and rates of gene flow. These are very difficult to estimate for *Drosophila*. Contrary to the effects of selection, drift acts uniformly over all loci within the genome; and except for sampling variation, the variance of gene frequencies should be homogeneous under the action of genetic drift. The results from *D. affinis* and *D. persimilis* (Taylor and Powell, 1977) do not support the hypothesis that genetic drift is primarily responsible for the differentiation in gene frequencies which is observed. This is for two reasons. Not all loci studied show the significant spatial heterogeneities which would be predicted by a genetic drift model. One might argue that allozyme loci which have homogeneous gene frequency distributions over the areas studied are acted on by selection, but those showing heterogeneities are not. Chromosome inversions which contain many loci should therefore show homogeneous spatial distributions. The three studies of microspatial heterogeneity in inversion frequencies summarized in Table 1 all report the discovery of microgeographic differences. A second reason to discard the genetic drift hypothesis is provided by data on dispersal in *Drosophila* of which *D. persimilis* is best studied (Powell et al., 1976). *D. persimilis* is capable of dispersing several hundred meters a day, and preliminary data on *D. affinis* show that about one-quarter of a population of marked flies dispersed at least 75 m in 24 hours. Begon (1976) has recently published estimates of both dispersal rates and population sizes in *D. subobscura* which suggest that this species is capable of moving about 50 m on the average each day and that effective population sizes are on the order of 10,000. All these results suggest that genetic drift is not a factor likely to be significant in the maintenance of microspatial differences in gene frequencies. It is important to issue a caveat at this point before the importance of genetic drift is discounted. The estimation of gene flow from dispersal data can be misleading (Ehrlich and Raven, 1969). Although they will be difficult to determine, estimates of gene flow for *Drosophila* populations are sorely needed if the mechanisms leading to population structuring are to be fully understood.

2. Natural Selection

If the pattern of selection within the environment occupied by a panmictic population is a mosaic one, i.e. different genotypes favored in different patches, the spatial differentiation observed may be due to selection which acts in each generation to reduce the frequencies of less fit genotypes in each patch. How much differentiation resulted from selection would depend upon the degree of movement by flies, and the spatial arrangement and frequency of the different habitats. Taylor (Taylor and Powell, 1977) has presented a model in which the rate of change of gene frequencies in an environment containing two alternating habitat types is represented by the following equations:

$$\Delta p_1 = (1-m)\frac{p_1(1-p_1)}{1-s_1p_1^2-s_2(1-p_1)^2}[s_2-p_1(s_1+s_2)] + m(p_2-p_1)$$

$$\nabla p_2 = (1-m)\frac{p_2(1-p_2)}{1-s_3p_2^2-s_4(1-p_2)^2}[s_3-p_2(s_3+s_4)] + m(p_1-p_2)$$

where m = migration rate between the patches,

p_1, p_2 = gene frequencies in patch 1 and 2 respectively,

s_1, s_2 = selection coefficients against the two homozygous genotypes in patch 1

s_3, s_4 = selection coefficients against the two homozygous genotypes in patch 2.

This model assumes heterotic selection and two alleles at a locus. Using an estimate of m = 33% exchange between patches in each generation, Taylor finds that selection coefficients of 50% result in only 4% gene frequency differences between patches. This difference is less than the 5.4% difference in inversion frequencies between sites observed by Taylor and Powell. Thus in order to explain the difference in gene frequencies observed in various studies, selection coefficients in excess of 50% would have to be operational. Such strong selection seems most unlikely.

The data from the Est-6 locus of *D. affinis* suggest that the frequency of the predominant allele at this locus varies clinally. In recent years a number of both theoretical and experimental studies of clinal variation have appeared. Endler (1973) has investigated the effects of gene flow between adjacent demes on the maintenance of a cline in gene frequencies. In experiments utilizing both experimental *Drosophila* populations and computer simulation, Endler showed that population differentiation is easily maintained even for 100% gene flow. Moreover, Endler's analysis shows that even a weak environmental gradient may result in spatial differentiation along a series of demes each exchanging migrants with their immediate neighbors. Taylor and Powell (1977) used a model of Slatkin's (1973) relating the rate of gene frequency change along a linear cline as a function of selection coefficients, degree of dominance, and the standard deviation of dispersal distances to estimate selection coefficients maintaining spatial differences in inversion frequencies. They found that inversion frequency differences of 5.4% between habitats separated by 335 m could be accounted for by selection coefficients in the range of 0.05 - 0.10. Clearly these two sets of analyses, one based on an "alternating patch" model and the second based on a linear cline yield conflicting results. If assumptions about gene flow are correct, selection would need to be very strong to maintain spatial differentiation in an alternating patch model. Clinal models result in selection coefficients which are quite reasonable, but this model does not apply to all spatial variation observed and makes the unrealistic assumption that spatial variation occurs in only one dimension.

The degree of gene flow between demes or along a cline is a critical parameter in any attempt to determine the magnitude of selection coefficients. Data on dispersal rates are not available for *D. affinis*, but data are available for *D. persimilis* (Powell et al., 1976), *aldrichi* (Richardson, 1969), *subobscura* (Begon, 1976) and other species. Estimates of the standard deviation of dispersal rates range from 25 m per generation in *D. aldrichi* to 561 m in *D. persimilis*. For the cline in the frequency of the Est-6^{102} allele in *D. affinis*, these figures result in an estimated selection coefficient ranging from 0.006 to 2.91 using Slatkin's equation for a linear cline. While the available data do not rule out the possibility that microspatial variation in gene frequencies is maintained by selection, the most reasonable model predicts selection coefficients which are larger than any thought to operate under natural conditions.

3. Non-uniform Dispersal

Collections of *D. affinis* within the sample grid described previously and in recent experiments described below show that one area of the grid tends to have a higher density than other areas. This high density area may represent the most suitable habitat for *D. affinis*. If the different genotypes studied have different dispersal rates, then the most active genotypes would tend to move farthest from areas of high density and thus create spatial differences in gene frequencies. Begon's (1976) finding that there are sex differences in dispersal rates in *D. subobscura* makes this hypothesis reasonable. However, our early results on habitat selection in *D. affinis* make the hypothesis unlikely.

4. Habitat Selection

Thorpe (1945) drew attention to habitat selection as a behavioral mechanism which was likely to lead to spatial isolation and which could work in concert with geographical separation as a means of developing reproductive isolation. Other workers (Maynard Smith, 1966; Gillespie, 1974; Christiansen, 1975; Taylor, 1975, 1976) have shown that habitat selection can act to maintain genetic polymorphisms without the requirement of marginal overdominance. The data we have obtained on *D. affinis* and that available for *melanogaster*, *aldrichi*, *subobscura*, and *persimilis* could be explained if flies of different genotypes showed different habitat preferences. In their study of *D. persimilis* Taylor and Powell (1977) conclude that habitat selection is the most likely explanation for the spatial differentiation in gene and inversion frequencies they observed. These investigators have since obtained independent data which support their contention.

We have studied habitat selection in *D. affinis* by simultaneously collecting flies from the two ends of the grid described above. One-thousand five hundred flies were marked using flourescent pigments and immediately released at the center of the grid. After 24 hours, 2,556 flies were collected in the upper and lower areas of the grid and examined for the presence of a mark. Of 227 marked and recaptured flies initially collected in the lower area of the grid, 148 or 65.2% were recaptured there 24 hours later. Of 125 marked and recaptured flies initially captured in the upper area of the grid, 52 or 41.6% were recaptured there. These figures are not significantly different from random choice, and this preliminary experiment failed to support the hypothesis that habitat selection is playing a role in maintaining the observed spatial differentiation in *D. affinis*.

D. Conclusion

As with many other animal and plant species studied (Hedrick et al., 1976), *Drosophila* populations may show spatial differentiation of gene frequencies. This variation in gene frequencies, often restricted to areas on the order of several thousand square meters, can explain much of the variation found over wide geographic areas. This conclusion is demonstrated in Table 5 for the Est-6 locus in *D. affinis*.

Table 5. Comparison of gene frequency differences within a local population to those present on a geographic scale for *Drosophila affinis*

Variance Component	Unionville Grid	Eastern U.S.
σ_T^2 = Total variance at Est-6 locus	0.01939	0.00815
σ_E^2 = Sampling variance for Est-6^{102} allele	0.00462	0.00116

Average differences in frequency of Est-6^{102} among 13 populations = 0.116

$$\text{Correction for sampling error} = \frac{\sigma_T^2 - \sigma_E^2}{\sigma_T^2}$$

$$\text{Corrected average difference} = (0.116)\ \frac{(0.00815-0.00116)}{0.00815} = 0.100$$

Average difference in frequency of Est-6^{102} between the ends of the Unionville Grid = 0.122

$$\text{Corrected average difference} = 0.122\ \frac{(0.01939-0.00462)}{0.01939} = 0.093$$

The total variance in gene frequencies was calculated as

$$\Delta_T^2 = \sum_i \sum_r \frac{N_r}{N} (p_{ir} - \bar{p}_i)^2$$

where: N_r number of individuals in the rth population,

N = total number of individuals in all populations,

p_{ir} = frequency of the ith allele in the rth population, and

$\bar{p}_i$ = mean frequency of the ith allele in all populations.

The sampling variance for the Est-6^{102} allele was estimated as

$$\frac{\bar{p}(1-\bar{p})}{\bar{N}} \quad \text{where:}$$

$\bar{p}$ = mean frequency of Est-6^{102} in all populations, and

$\bar{N}$ = mean number of genes sampled per population.

Estimates of the frequency of the Est-6^{102} allele are available for 13 geographic populations spread throughout the eastern U.S. (Richmond, unpublished results). The mean difference in the frequency of the allele over 78 possible comparisons among 13 populations is 0.116. Correction of this figure for sampling error yields an average difference of 0.100 (Table 5). A similar analysis of gene frequencies from the Unionville grid (Tables 3, & 5) results in an average difference between the upper and lower areas of 0.093. At least for this locus, small scale microgeographic variation can explain practically all the geographic variation in gene frequencies. This result supports the hypothesis of Levins (1968) that selective factors acting on *Drosophila* may be quite similar over wide geographic areas.

The theory that selective processes in a natural population will be most effective if the population is subdivided into a series of demes was advanced long ago by Sewall Wright (1929). He realized that such a subdivided population would allow for a continuing series of evolutionary experiments, and in a sense allow the composite population to be preadapted to future environmental change. Wright's model was attractive, but many evolutionary biologists discounted its significance because it was generally believed that for mobile species gene flow between demes would prevent any spatial differentiation of gene frequencies. In a seminal paper, Ehrlich and Raven (1969) questioned this view and turned the attention of their readers to the possibilities for intrapopulation variation. The work from *Drosophila* summarized here presents indirect evidence that gene flow in some of these species may be relatively unimportant in maintaining the genetic continuity of species. A problem of major importance for those concerned with the evolution of *Drosophila* populations is to obtain direct estimates of gene flow within and between populations. The demonstration that individuals are capable of moving over wide areas is clearly not sufficient to demonstrate gene flow, only that it may occur. A full explanation for the observed microgeographic differences in gene frequencies in *Drosophila* depends upon an estimation of the significance of gene flow.

E. Summary

The possible significance of population structure in the evolution of species was recognized early in the development of population genetic theory. However, the view that the genetic structure of species is maintained by gene flow among panmictic populations directed the attention of evolutionary biologists to large scale geographic variation in gene frequencies rather than to the analysis of microgeographic variation within populations. Studies of several *Drosophila* species reveal that these organisms are capable of responding to small scale environmental changes and that such environmental variation can affect the maintenance of genetic variation within artificial populations. Field studies of microspatial variation in *Drosophila* have revealed that significant differences in chromosome inversion frequencies and allozyme-gene frequencies are the rule rather than the exception. Preliminary data suggest that behavioral habitat selection may be important in maintaining microspatial differences in gene frequencies in some species.

Acknowledgements. Many persons have contributed freely of their ideas and time to the elaboration of the data on *Drosophila affinis* which is reported here. I am particularly thankful for the help provided by Michael D. Sabath, J. Michael Jones, Anna Finkel, Dwight D. Miller, Max Levitan, Mark Gromko, Bruce Cochrane, Donald Pyle and Ann W. Richmond. Our research has been generously supported by grants from the National Institutes of Health.

References

Beardmore, J.: Ecological factors and the variability of gene pools in *Drosophila*. In Essays in Evolution and Genetics in Honor of Theodosius Dobzhansky, (ed. M.K. Hecht and W.C. Steere). New York: Appleton-Century-Crofts, 1970, pp. 299-314.

Begon, M.: Dispersal density and microdistribution in *Drosophila subobscura* Collin. J. Animal Ecol. 45:441-456. (1976).

Carson, H.L.: Three flies and three islands: Parallel evolution in *Drosophila*. Proc. Natl. Acad. Sci. U.S.A. 71:3517-3521. (1974).

Cavalli-Sforza, L.L., Bodmer, W.F.: The Genetics of Human Populations. San Francisco: W.H. Freeman and Co., 1971.

Christiansen, F.B.: Hard and soft selection in a subdivided population. Amer. Natur. 109:11-16. (1975).

Crow, J.F., Kimura, M.: An Introduction to Population Genetics Theory. New York: Harper and Row, 1970.

Crumpacker, D.W.: The use of micronized fluorescent dusts to mark adult *Drosophila pseudoobscura*. Am. Midl. Nat. 91:118-129. (1974).

Dempster, E.R.: Maintenance of genetic heterogeneity. Cold Spring Harbor Symp. Quant. Biol. 20:25-32. (1955).

Dobzhansky, T.: A review of some fundamental concepts and problems of population genetics. Cold Spring Harbor Symp. Quant. Biol. 20:1-15. (1955).

Ehrlich, P.R., Raven, P.H.: Differentiation of populations. Science 165:1228-1232. (1969).

Endler, J.A.: Gene flow and population differentiation. Science 179:243-250. (1973).

Gillespie, J.: Polymorphism in patchy environments. Amer. Natur. 108:145-151. (1974).
Hedrick, P.W., Ginevan, M.E., Ewing, E.P.: Genetic polymorphism in heterogeneous environments. Ann. Rev. Ecol. Syst. 7:1-32. (1976).
Heed, W.B.: Host plant specificity and speciation in Hawaiian *Drosophila*. Taxon 20:115-121. (1971).
Johnston, J.S., Heed, W.B.: Dispersal of desert-adapted *Drosophila:* The saguaro-breeding *D. nigrospiracula*. Amer. Natur. 110:629-651. (1976).
Kaneshiro, K.Y., Carson, H.L., Clayton, F.E., Heed, W.B.: Niche separation in a pair of homosequential *Drosophila* species from the island of Hawaii. Amer. Natur. 107:766-774. (1973).
Krimbas, C.B., Alevizos, V.: The genetics of *Drosophila subobscura* IV. Further data on inversion of polymorphism in Greece--Evidence of microdifferentiation. Egyp. J. Genet. Cytol. 2:121-132. (1973).
Levene, H.: Genetic equilibrium when more than one ecological niche is available. Amer. Natur. 87:331-333. (1953).
Levins, R.: Evolution in Changing Environments. Princeton, New Jersey: Princeton University Press, 1968.
Li, C.C.: Population Genetics. Chicago: Univ. of Chicago Press, 1955.
Maynard Smith, J.: Sympatric speciation. Amer. Natur. 100:637-650. (1966).
Mayr, E.: Animal Species and Evolution. Cambridge, Mass.: Harvard University Press, 1963.
McDonald, J.F., Ayala, F.J.: Genetic response to environmental heterogeneity. Nature 250:572-574. (1974).
McKenzie, J.A., Parsons, P.A.: Alcohol tolerance: An ecological parameter in the relative success of *Drosophila melanogaster* and *Drosophila simulans*. Oecologia 10:373-388. (1972).
McKenzie, J.A., Parsons, P.A.: Microdifferentiation in a natural population of *Drosophila melanogaster* to alcohol in the environment. Genetics 77:385-394. (1974).
Miller, D.W.: Geographical distributions of the American *Drosophila affinis* subgroup species. Am. Midl. Natur. 60:52-70. (1958).
Powell, J.R.: Genetic polymorphisms in varied environments. Science 174:1035-1036. (1971).
Powell, J.R., Dobzhansky, T., Hook, J., Wistrand, H.: Genetics of natural populations. XLIII. Further studies on rates of dispersal of *Drosophila pseudoobscura* and its relatives. Genetics 82:493-506. (1976).
Richardson, R.H.: Migration, and isozyme polymorphisms in natural populations of *Drosophila*. Proc. XII Int'l Congr. Genet. II. 155-156. (1968).
Richardson, R.H.: Migration, and enzyme polymorphisms in natural populations of *Drosophila*. Jap. J. Genet. 44:(Suppl. 1) 172-179. (1969).
Richardson, R.H., Johnston, J.S.: Behavioral components of dispersal in *Drosophila mimica*. Oecologia 20:287-299. (1975a).
Richardson, R.H., Johnston, J.S.: Ecological specialization in Hawaiian *Drosophila*. Habitat selection in Kipuka Puaulu. Oecologia 21:193-204. (1975b).
Richmond, R.C.: Microgeographic genetic differentiation in a natural population of *Drosophila affinis*. (MS, 1977).
Rockwell, R.F., Sieger, M.B.: Phototaxis in *Drosophila:* A critical evaluation. Am. Sci. 61:339-345. (1973).
Rockwood-Sluss, E.S., Johnston, J.S., Heed, W.B.: Allozyme genotype-environment relationships. I. Variation in natural populations of *Drosophila pachea*. Genetics 73:135-146. (1973).
Slatkin, M.: Gene flow and selection in a cline. Genetics 75:733-756. (1973).

Stalker, H.D.: Chromosome studies in wild populations of *D. melanogaster*. Genetics 82:323-347. (1976).

Taylor, C.E.: Genetic loads on heterogeneous environments. Genetics 80:621-635. (1975).

Taylor, C.E.: Genetic variation in heterogeneous environments. Genetics 83:887-794. (1976).

Taylor, C.E., Powell, J.R.: Microgeographic differentiation of chromosomal and enzyme polymorphisms in *Drosophila persimilis*. Genetics 85:681-695. (1977).

Thorpe, W.H.: The evolutionary significance of habitat selection. J. Anim. Ecol. 14:67-70. (1945).

Waddington, C.H., Woolf, B., Perry, M.M.: Environment selection by *Drosophila* mutants. Evolution 8:89-96. (1954).

Wallace, B.: Distance and the allelism of lethals in a tropical population of *Drosophila melanogaster*. Amer. Natur. 100:565-578. (1966).

Wallace, B.: Observations on the microdispersion of *Drosophila melanogaster*. In Essays in Evolution and Genetics in Honor of Theodosius Dobzhansky. (ed. M.K. Hecht and W.S. Steere). New York: Appleton-Century-Crofts, 1970, pp. 381-389.

Wright, S.: Evolution in a Mendelian population. Anat. Rec. 44:287. (1929).

4 Other Animals

Ecological Parameters and Speciation in Field Crickets

R.G. Harrison

A. Introduction

The study of animal speciation has traditionally focused on spatial relationships among diverging populations. The widely accepted model of allopatric speciation (Mayr, 1963a; Dobzhansky, 1970) suggests that new species arise in geographically isolated populations through the gradual accumulation of genetic differences. Mayr (1963a) contends that geographic speciation is "the almost exclusive mode of speciation among animals" (p. 481). However, detailed ecological, behavioral, and cytogenetic studies of a variety of insect taxa provide evidence that other modes of speciation commonly occur (see White, 1974; Bush, 1975a for reviews). The evolution of reproductive isolation may often proceed rapidly and involve changes at relatively few genetic loci. Intrinsic, as well as extrinsic, barriers to gene flow are potentially important in the speciation process.

In order to explain life cycle differences between sibling species of field crickets, Alexander and Bigelow (1960) suggested that reproductive isolation may arise through temporal isolation of previously interbreeding populations. They termed this process "allochronic speciation." More recently, Alexander (1968) has provided additional arguments in support of this model. Here, I reexamine the evolution of life cycle differences in field crickets and attempt to evaluate the role of temporal isolation in the speciation process.

In many groups of animals and plants life cycles and development rates differ significantly among closely related species and among conspecific populations. Nowhere is this better illustrated than among the insects. Differences in number of generations per year, diapause stage, diapause intensity, termination of diapause, length of adult life, and duration of nymphal development are common, although certain taxa appear more variable than others (Lees, 1955; Masaki, 1961, 1965; Danilevskii, 1965; Alexander, 1968).

We may distinguish at least four factors that will influence the evolution of insect life cycles.

1. Adaptation to climate, particularly temperature. This may involve avoidance of unfavorable conditions through diapause or taking advantage of the full growing season through adjustment of development rate.

2. Dependence on a resource which is available only part of the year or for which quality changes with season.

3. Avoidance of predators, parasites, or competitors which occur seasonally.

4. Selection for pre-mating isolation (different breeding seasons) in areas of overlap between populations/species that produce hybrids of reduced viability or fertility.

By examining groups of closely related species which exhibit a variety of life cycle adaptations, we may ascertain which of these selective forces has been important and whether life cycle shifts have been instrumental in the speciation process.

B. Life Cycles in Two Genera of Field Crickets

Field crickets in the genera *Gryllus* and *Teleogryllus* are excellent material for the study of speciation. These crickets are large, hemimetabolous insects, widely distributed throughout temperate and tropical regions. There is no evidence in the literature to suggest that they depend on seasonally available resources. Species of *Gryllus* are omnivorous. They have been reported as pests of strawberries (Thomas and Reed, 1937), cotton (McGregor, 1929; Folsom and Woke, 1939), and alfalfa grown for seed (Severin, 1935), as predators of grasshopper eggs (Criddle, 1925; Smith, 1959) and of the pupae of the apple maggot *Rhagoletis pomonella* (Monteith, 1971); they also appear to feed on carrion and a variety of plant material (Severin, 1935; Cantrall, 1943). Information about levels of predation and parasitism is not available for most species (but see Severin, 1935 and Cade, 1975 for limited data on levels of parasitism in *Gryllus pennsylvanicus* and *G. integer*). It is possible that predation pressure may vary seasonally or that escape in time is a successful strategy for avoiding parasites. However, other evidence strongly suggests that adaptation to climate has been the determining factor in the evolution of life cycle differences in these crickets (Alexander, 1968).

The well-studied groups of *Gryllus* and *Teleogryllus* are shown in Figure 1. There are four tropical species (*Gryllus assimilis*, *Gryllus bimaculatus*, *Teleogryllus mitratus*, and *Teleogryllus oceanicus*), each of which has one or more near relatives in the adjacent temperate zone. These crickets exhibit four distinct types of life cycles (Fig. 2). In the tropics species may breed throughout the year, although the relative abundance of particular developmental stages will vary. Generations are not discrete. In the temperate regions species of *Gryllus* and *Teleogryllus* are either univoltine or bivoltine, and diapause occurs either in the egg stage or in a late nymphal instar. Similar life cycles have apparently evolved independently in the two genera. If temperate species are derived from tropical ancestors, then in each case this transition has been accompanied by a shift in life cycle -- the evolution of diapause. Transitions from univoltine or bivoltine life cycles to continuous breeding are also theoretically possible (Alexander, 1968).

Speciation in field crickets has generally not involved major changes in morphology, chromosome number, or electrophoretic mobility of soluble proteins. *Gryllus assimilis*, *Teleogryllus mitratus*, and *Teleogryllus oceanicus* are all exceedingly similar morphologically to their temperate relatives. In each group the species were originally distinguished on the basis of differences in calling song, life cycle or habitat (Ohmachi and Matsuura, 1951; Fulton, 1952; Alexander, 1957; Bigelow and Cochaux, 1962; Leroy, 1963; Hogan, 1965). Only subsequently have diagnostic morphological characters been discovered. All *Gryllus* species have a male diploid chromosome number of 29 (Randell and Kevan, 1962; Lim et al., 1973). *Teleogryllus* species have a male diploid chromosome number of 27, except *T. yezoemma*, for which 2n♂ = 25 (Ohmachi and Masaki, 1964; Fontana and Hogan, 1969; Lim et al., 1969). Based on electrophoretic comparisons at 21 loci coding for soluble

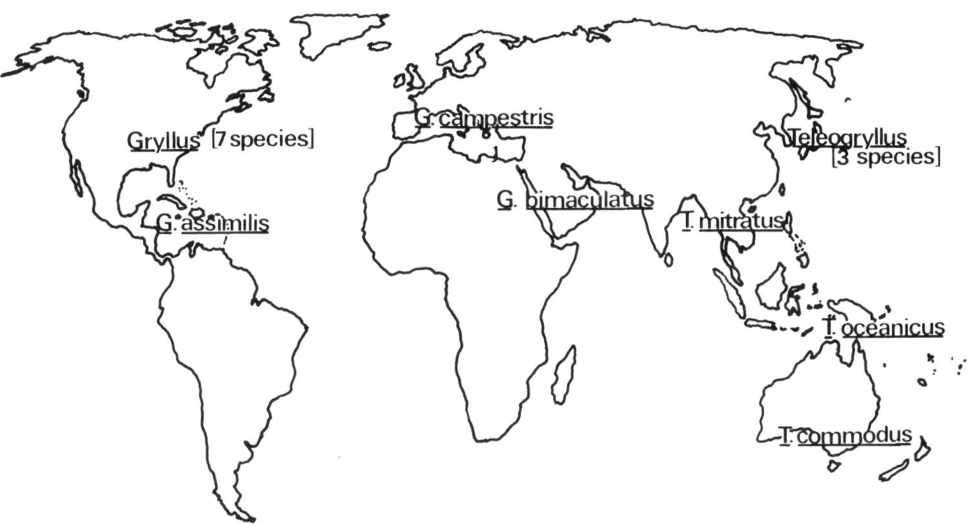

Fig. 1. Closely related pairs or groups of species in the genera *Gryllus* and *Teleogryllus*. Only well-studied species are shown. Distributional limits are not accurately known in certain cases.

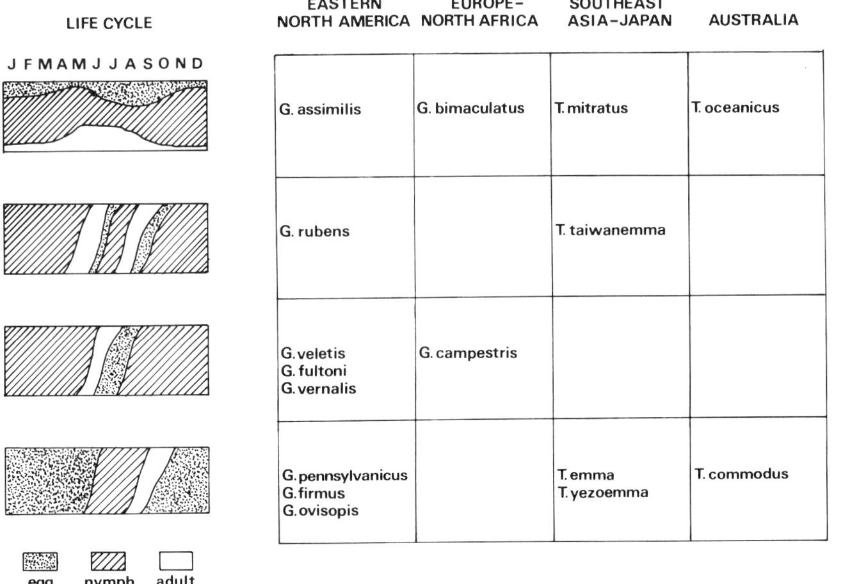

LIFE CYCLE
J F M A M J J A S O N D
EASTERN NORTH AMERICA
EUROPE-NORTH AFRICA
SOUTHEAST ASIA-JAPAN
AUSTRALIA
G. assimilis
G. bimaculatus
T. mitratus
T. oceanicus
G. rubens
T. taiwanemma
G. veletis
G. fultoni
G. vernalis
G. campestris
G. pennsylvanicus
G. firmus
G. ovisopis
T. emma
T. yezoemma
T. commodus
egg
nymph
adult

Fig. 2. Life cycles of temperate and tropical field crickets in the genera *Gryllus* and *Teleogryllus*. Considerable variation (in development rates, time of hatching, sensitivity to environmental cues) exists among species with the same fundamental developmental pattern.

proteins, the genetic distance (Nei, 1972) between *T. oceanicus* and *T. commodus* is 0.09 (Harrison, unpublished). This is a relatively small difference, even for sibling species.

In Australia and around the Mediterranean Sea, non-diapausing and diapausing species overlap without interbreeding. In neither case are post-mating reproductive isolating mechanisms well developed. Hybrids from *G. bimaculatus* X *G. campestris* are viable and fertile (Cousin, 1958; Grellet, 1961). In the *T. oceanicus* X *T. commodus* cross only male hybrids are fertile, but backcross progeny are easily produced (Leroy, 1966; Hoy, personal communication). The isolating mechanisms operating in the field appear to be different in the two cases. Adults of *T. oceanicus* and *T. commodus* are sympatric and synchronic over a wide area, but differences in the species-specific male calling song (both in frequency and temporal pattern) appear to effectively isolate the two species. Hybrid song, which is distinctive, is not heard in the area of sympatry (Hill et al., 1972). *G. bimaculatus* and *G. campestris*, on the other hand, have nearly identical calling songs (Leroy, 1966; Alexander, 1967), but seem to occupy different habitats, *campestris* preferring drier areas (Finot, 1883; Chopard, 1922).

From these examples it is clear that the evolution of diapause is not sufficient to isolate continuously breeding tropical species from their temperate relatives. A likely scenario for speciation in these situations would involve isolation during the Pleistocene of populations at the northern or southern margin of a continuously breeding species, selection for diapause in these isolates during cold intervals, and eventual reestablishment of contact. Hybrids from crosses between non-diapausing and diapausing species exhibit little or no tendency to diapause (Bigelow 1960a, 1962; Bigelow and Cochaux, 1962; Ohmachi and Masaki, 1964; Hogan, 1965; Masaki and Ohmachi, 1967; MacFarlane and Drummond, 1970). Therefore, selection for pre-mating isolating mechanisms would be strong among diapausing individuals in areas with cold winters.

The obvious selective advantage of diapause is that it insures that all individuals will be in a cold-resistant stage at the onset of winter. Diapause, however, also serves to synchronize the appearance of adults and thus permits displacement in time of breeding populations.

C. Evolution of Diapause in North American Field Crickets

In eastern North America, field crickets have been described with each of the four life cycles (Fig. 2). Additional information on life cycle, distribution, and habitat is summarized in Table 1.

Of particular interest are the sibling species *Gryllus veletis* and *G. pennsylvanicus*, which are sympatric throughout the northeastern United States, occupy the same habitats, are exceedingly similar morphologically, and have identical calling songs. They were separated only on the basis of differences in their life cycles (Alexander and Bigelow, 1960). *G. veletis* diapauses as a late instar nymph, and adults are present from May until July. *G. pennsylvanicus* diapauses in the egg stage and matures in August. In both species diapause is obligate; developmental patterns are genetically determined, not environmentally conditioned (Bigelow, 1958, 1962; Rakshpal, 1962). Although there may be a brief period in mid-summer when adults of the two species overlap (Alexander and Meral, 1967), hybridization does

Table 1. Distribution, habitat and life cycle in eastern North American species of *Gryllus*

Species	Distribution	Habitat	Adults present
veletis	Northern U.S. and southern Canada	Disturbed areas: pastures, fields, roadsides, lawns	May-July
pennsylvanicus	Northern U.S. and southern Canada	Disturbed areas: pastures, fields, roadsides, lawns	Aug.-Oct.
rubens	Southeastern U.S.	Disturbed areas: pastures, fields roadsides, lawns	Mar.-June July-Sept.
firmus	Coastal areas of the southeastern U.S.	Flat sandy areas bordering ocean beaches; sandy areas inland in Fla., N.C.	Aug.-Oct. (May-July)
fultoni	Southeastern U.S.	Leaf litter in deciduous and pine-deciduous forests, forest borders	April-July (throughout the year, except Oct., Nov., in Florida)
vernalis	East central U.S.	Leaf litter in deciduous forests	April-July
ovisopis	Northern Florida	Leaf litter, under logs in broadleaf forests	Sept.-Dec.
assimilis	South Florida	Disturbed areas, grasslands	Throughout the year

From Fulton, 1952; Alexander, 1957, 1968; Alexander and Walker, 1962; and Walker, 1974.

not occur. Moreover, despite repeated attempts, hybrid progeny have never been produced in laboratory matings (Bigelow, 1958, 1960a; Alexander and Bigelow, 1960).

In order to explain the similarity in morphology, song, habitat, and distribution of *G. veletis* and *G. pennsylvanicus* and the difference in their life cycles, Alexander and Bigelow (1960) proposed the concept of allochronic speciation. This would involve (1) elimination during winters of all developmental stages except eggs and late instar nymphs, resulting in two temporally isolated populations; (2) geographic isolation of these populations from others (in warmer climates) in which breeding is continuous and all stages are found throughout the year (see Alexander, 1963; Mayr, 1963b for discussion); (3) evolution of obligate diapause in the two overwintering stages, thus assuring continued temporal isolation; (4) appearance of reproductive incompatability. Steps (3) and (4) may be inseparable; i.e., reproductive incompatability may be a direct result of the evolution of life cycle differences. That hybrids are never produced from crosses between univoltine egg and nymphal diapausing species would seem to support this hypothesis (Alexander, 1968).

Also important in this reconstruction are observations on *G. firmus*, a species found along the south Atlantic or Gulf Coasts of the United States. Most individuals of *G. firmus* have life cycles similar to *G. pennsylvanicus*; diapause occurs in the egg stage; and adults appear in late summer and early fall. However, small populations of spring adults are also found in certain localities, where they are sympatric with larger fall adult populations (Fulton, 1952; Alexander, 1968). Alexander and Bigelow (1960) suggest that this situation may represent an early stage in the same process of allochronic speciation.

Estimates of genetic distance between species, based on comparison of soluble proteins by gel electrophoresis, are not consistent with this model (Harrison, 1977). From a matrix of genetic distances for five eastern species and two populations from the southwestern United States, I have constructed an evolutionary tree for North American *Gryllus* (Fig. 3) using the Wagner tree method of Farris (1972), which produces a minimum length tree that allows for heterogeneous rates of evolution.

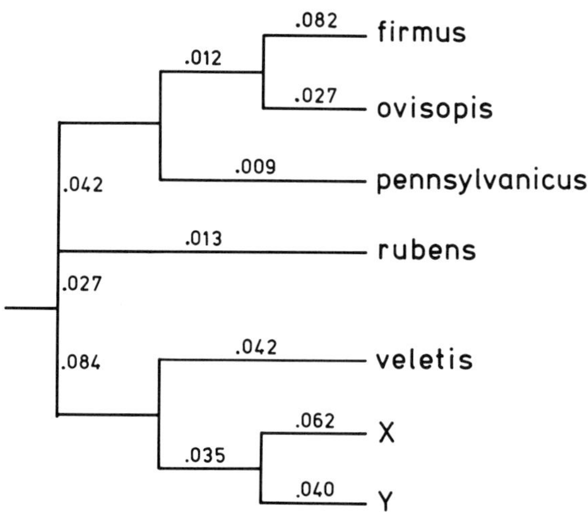

Fig. 3. Evolutionary tree of North American *Gryllus* species. X and Y are populations from southern Utah and northern New Mexico, respectively. The numbers represent the presumed amount of protein evolution that has occurred in each branch, in units of genetic distance (Nei, 1972). See text for further details.

The following points should be noted from examining Fig. 3: (1) *G. veletis* and *G. pennsylvanicus* do not appear closely related. Indeed, the genetic distance between these two species (0.17) is greater than that for many other pairs in this genus (Harrison, 1977). (2) *G. pennsylvanicus*, *G. firmus*, and *G. ovisopis* form one distinct lineage. These are the three egg-diapausing species. (3) *G. veletis* forms a second lineage with the two populations from the southwestern United States. These are all apparently univoltine, nymphal-diapausing species. (4) *G. rubens*, a bivoltine species, appears to represent a third lineage.

The major conclusion is that each temperate life cycle has arisen only once in the evolution of the genus in North America. Separation of egg and nymphal diapausing crickets probably occurred relatively early in this evolutionary history.

If these phylogenetic relationships derived from electrophoretic data are correct, then the hypothesis of allochronic speciation for *G. veletis* and *G. pennsylvanicus* is no longer a particularly attractive one. It is more reasonable to assume that life cycles involving egg diapause and nymphal diapause appeared independently in allopatric populations, perhaps at different times during the Pleistocene. The existence of off-cycle individuals of *G. firmus* may best be explained by premature hatching of eggs in late summer and survival of nymphs during the winter. There is no indication that spring adults have evolved any tendency to diapause as nymphs.

The similarity in song of *G. veletis* and *G. pennsylvanicus* must reflect parallel evolution or, more likely, that calling song is unchanged from an ancestral *Gryllus*. It suggests that adults of the two species have never overlapped both in time and space. In general, allopatric or allochronic species may have the same song; sympatric and synchronic species are likely to have different song patterns as the result of selection for premating isolation (Alexander, 1962, 1967).

I might also point out here that if genes affecting male calling songs also affect female song preference, then song differences could arise through genetic drift in small, isolated populations (in a manner analagous to that suggested for chromosomal rearrangements (Bush, 1975a). Hoy et al., (1977) have evidence that such genetic coupling may exist.

D. Development Rates, Adaptation, and Genetic Differentiation

Although the variety of diapause strategies is certainly the most striking life cycle characteristic of North American *Gryllus*, differences in development rate between populations with the same life cycle may also be important for speciation. The three egg-diapausing species in eastern North America have very different development times in the field (Table 2). Development times are longer for

Table 2. Approximate time from hatching to adult in three egg-diapausing species of *Gryllus* from the eastern United States. Data are from crickets in natural populations and from individuals reared in outdoor cages.

Species	Latitude	Time from hatching to adult (days)	Reference
ovisopis	30°	110	Walker 1974
firmus	34°	90	Fulton 1952
pennsylvanicus	42°	65	Harrison 1977

species from warmer areas. Similar trends are observed within species. Thus, Masaki (1967) has shown for the Japanese field cricket, *Teleogryllus emma*, that if crickets from a series of populations are brought into the laboratory and reared under identical conditions, development time is negatively correlated with latitude (Fig. 4). Crickets from northern areas develop more rapidly and give rise to smaller adults. A similar pattern is also evident from comparisons of development times in *G. pennsylvanicus* populations (Fig. 5).

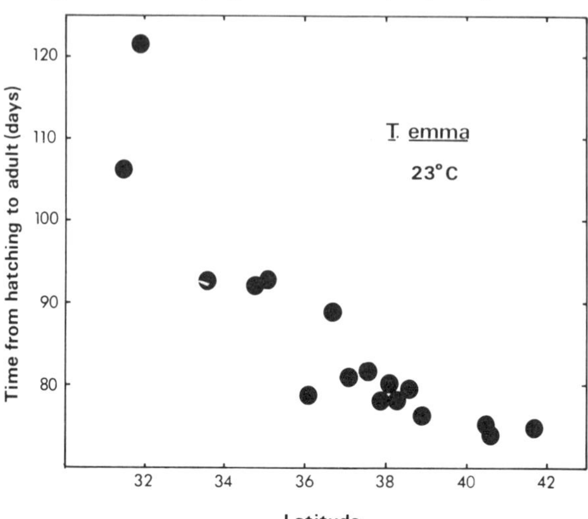

Fig. 4. Duration of nymphal development in laboratory reared populations of *Teleogryllus emma* collected from different localities. Data are from Masaki, 1967.

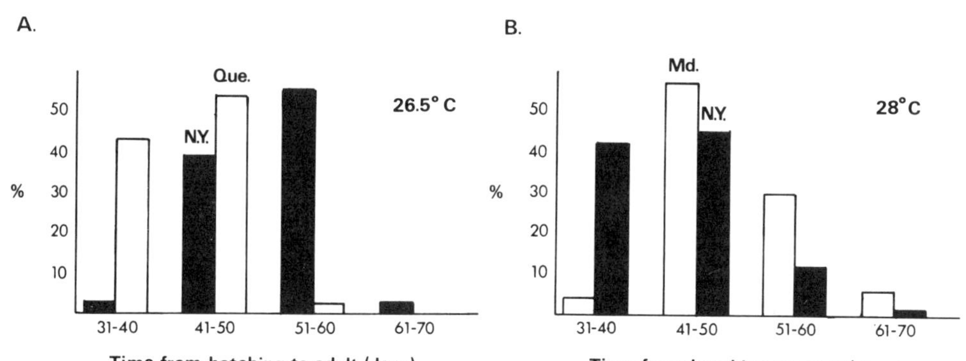

Fig. 5. Comparison of development times in laboratory reared *Gryllus pennsylvanicus*. (A) Populations from New York (42°) and Quebec (46°). Data from Bigelow, 1960b. (B) Populations from Maryland (39°) and New York (42°). Data from Harrison, 1977. In each case crickets from higher latitudes develop more rapidly. Note that the two experiments are not directly comparable because of difference in rearing conditions.

The variation in development almost certainly reflects adaptation to local environmental conditions (Masaki, 1967). In warmer areas crickets have a longer growing season. In univoltine species selection favors longer development times because (1) eggs laid too early risk breaking diapause before the onset of cold weather (this may occur in *G.firmus*) and (2) females that take longer to develop are larger as adults and have greater fecundity.

How does variation in development time affect the temporal relationships of conspecific populations? Is there any evidence that such differences provide barriers to gene flow?

Populations of *G. pennsylvanicus* from the northeastern United States exhibit striking heterogeneity in electromorph frequencies at several enzyme loci (Harrison, 1977). Ordination of these populations using principal components analysis (Fig. 6) indicates that Northern, Lowland, and Blue Ridge crickets each form a distinct cluster (see Figure legend for an explanation of the grouping of these populations).

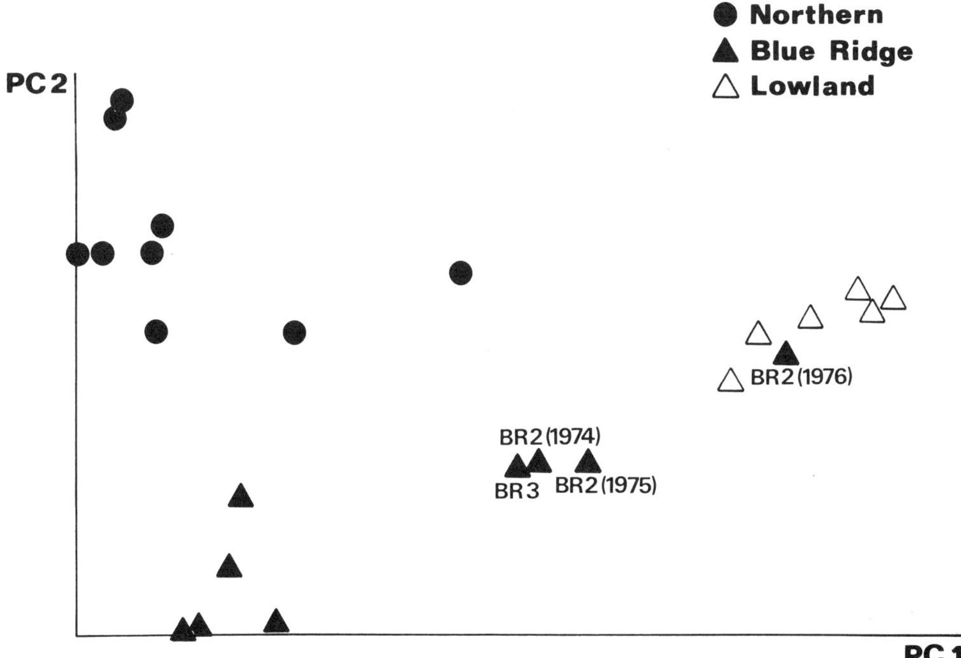

Fig. 6. Ordination of *Gryllus pennsylvanicus* populations by principal components analysis, based on electromorph frequencies at five loci (EST, GOT-1, GOT-2, MDH-2, PGI). Only the first two principal axes are shown, but these account for 80% of the variance. Northern populations are from New York, Connecticut, and northern Pennsylvania; Lowland populations from southern Pennsylvania, Maryland, and Virginia east of the Blue Ridge; Blue Ridge populations from points along the crest of the Blue Ridge in Virginia and North Carolina.

Crickets from two populations at the northern end of the Blue Ridge Parkway (BR2, BR3) appear intermediate between Lowland and Blue Ridge types. The BR2 and BR3 populations also differ from their neighbors on the Blue Ridge in duration of nymphal development. In general, crickets from the Blue Ridge become adult about two weeks earlier than crickets from Lowland areas. However, a significant fraction of the crickets collected at BR2 and BR3 have development times similar to those of Lowland crickets.

The BR2 and BR3 populations are located 30 miles west and south of Charlottesville, Virginia (a typical Lowland locality). Here the Blue Ridge rises abruptly 2000-3000 feet from the Piedmont area to the east. The eastern slope is almost entirely forested and constitutes a major barrier to dispersal for *G. pennsylvanicus*. Most individuals in this species have reduced hind wings and metathoracic flight muscles and are incapable of flight (Alexander, 1968; Harrison, 1977). Elsewhere (Harrison, 1977) I have suggested that Lowland and Blue Ridge populations represent locally adapted forms and that the anomalous populations on the Blue Ridge reflect recent colonization by crickets from Lowland areas, resulting in mixed populations.

Since Blue Ridge and Lowland crickets differ not only in electromorph frequency but also in development rate, we must ask whether the two forms interbreed at random where they occur together. I have compared electromorph frequencies in nymphs and adults collected on a single day, in each of three years. This provides a crude comparison between electromorph frequencies in crickets with slow and fast development rates (nymphs and adults, respectively). At the esterase locus, nymphs from BR2 have electromorph frequencies nearly identical to those in Lowland populations, whereas adults have electromorph frequencies similar to neighboring Blue Ridge populations (Harrison, 1977). In each of the three years there has been an excess of homozygotes at this same locus. These observations suggest that crickets from Lowland and Blue Ridge populations do not interbreed at random where they overlap or that hybrid offspring have reduced viability. Differences in development rate (resulting in different time of adult appearance) could contribute to inbreeding, since adults that appear early are likely to mate with each other. Evaluation of the significance of temporal isolation will require information on length of adult life, number of times each female mates during her lifetime, and the possibility of sperm storage and sperm precedence. In any event, it is evident that temporal isolation must play a more subtle role in genetic differentiation than that envisioned by Alexander and Bigelow (1960).

E. Conclusions

Seasonal isolation in closely related species of insects is not an uncommon phenomenon. Different development times result in temporal isolation of two "strains" of Diprionid sawflies feeding on balsam fir (Knerer and Atwood, 1973). Differences in photoperiod response appear to determine seasonal isolation of breeding adults of the green lacewings *Chrysopa carnea* and *C. downesi* (Tauber and Tauber, 1976; Tauber et al. 1977). In the genus *Rhagoletis*, Bush (1975b) has cited several examples of divergence in time of emergence between different host races. Among the Plecoptera (stoneflies) there are many examples of closely related species with seasonally isolated adults (Harper and Pilon, 1970; Nebeker, 1971; Hynes, 1976).

Seasonal isolation may provide an effective barrier to gene flow, but divergence in breeding time probably occurs primarily in allopatric populations as a result of local adaptation or in sympatric and parapatric populations that have diverged in resource utilization or habitat preference. However, adjustment of life cycles to local conditions may occur rapidly. For example, the European corn borer, introduced into the United States in this century, exhibits patterns of variation in diapause characteristics that appear to represent adaptation to latitudinal gradients (Beck and Apple, 1961; Sparks et al., 1966). Tauber et al. (1977) suggest that only two genes may be responsible for seasonal isolation of lacewing species. If life cycle changes occur rapidly, then barriers to gene flow need only persist for relatively short periods of time in order for significant divergence in development rate or breeding season to occur. In organisms of low vagility (such as many temperate species of field crickets) isolated populations must have occurred frequently during recent glacial periods.

Acknowledgements. I am particularly indebted to Peter Brussard and Ronald Hoy for providing advice and support during the course of my work on field crickets. I am also grateful to many other colleagues at Cornell University for helpful discussions.

References

Alexander, R.D.: The taxonomy of the field crickets of the eastern United States (Orthoptera:Gryllidae:*Acheta*). Ann. Ent. Soc. Amer. 50:584-602. (1957).

Alexander, R.D.: The role of behavioral study in cricket classification. Syst. Zool. 11:53-72. (1962).

Alexander, R.D.: Animal species, evolution, and geographic isolation. Syst. Zool. 12:202-204. (1963).

Alexander, R.D.: Acoustical communication in arthropods. Ann. Rev. Ent. 12:495-526. (1967).

Alexander, R.D.: Life cycle origins, speciation, and related phenomena in crickets. Quart. Rev. Biol. 43:1-41. (1968).

Alexander, R.D., Bigelow, R.S.: Allochronic speciation in field crickets and a new species, *Acheta veletis*. Evolution 14:334-346. (1960).

Alexander, R.D., Meral, G.: Seasonal and daily chirping cycles in the northern spring and fall crickets, *Gryllus veletis* and *G. pennsylvanicus*. Ohio J. Sci. 67:200-209. (1967).

Alexander, R.D., Walker, T.J.: Two introduced field crickets new to the eastern United States (Orthoptera:Gryllidae). Ann. Ent. Soc. Amer. 55:90-94. (1962).

Beck, S.D., Apple, J.W.: Effects of temperature and photoperiod on voltinism of geographic populations of the European Corn Borer, *Pyrausta nubilalis*. J. Econ. Ent. 54:550-558. (1961).

Bigelow, R.S.: Evolution in the field cricket, *Acheta assimilis* Fab. Can. J. Zool. 36:139-151. (1958).

Bigelow, R.S.: Interspecific hybrids and speciation in the genus *Acheta* (Orthoptera:Gryllidae). Can. J. Zool. 38:509-524. (1960a).

Bigelow, R.S.: Developmental rates and diapause in *Acheta pennsylvanicus* (Burmeister) and *Acheta veletis* (Alexander and Bigelow) (Orthoptera:Gryllidae). Can. J. Zool. 38:973-988. (1960b).

Bigelow, R.S.: Factors affecting developmental rates and diapause in field crickets. Evolution 16:396-406. (1962).
Bigelow, R.S., Cochaux, S.A.: Intersterility and diapause differences between geographic populations of *Teleogryllus commodus* (Walker) (Orthoptera:Gryllidae). Aust. J. Zool. 10:300-306. (1962).
Bush, G.L.: Modes of animal speciation. Ann. Rev. Ecol. Syst. 6: 339-364. (1975a).
Bush, G.L.: Sympatric speciation in phytophagous parasitic insects. In Evolutionary Strategies of Parasitic Insects and Mites (ed. P.W. Price). New York: Plenum, 1975b, pp. 187-206.
Cade, W.: Acoustically orienting parasitoids: Fly phonotaxis to cricket song. Science 190:1312-1313. (1975).
Cantrall, I.J.: The ecology of the Orthoptera and Dermaptera of the George Reserve. Misc. Publ. Mus. Zool., U. of Mich. No. 54: 1-182. (1943).
Chopard, L.: Faune de France. Orthoptères et Dermaptères. Paris: Paul Lechevalier, 1922.
Cousin, G.: Hybridations interspécifiques entre Gryllides et considérations sur l'évolution du groupe. Proc. X Int. Cong. Ent. 2: 881-897. (1958).
Criddle, N.: Field crickets in Manitoba. Can. Ent. 57:79-94. (1925).
Danilevskii, A.S.: Photoperiodism and Seasonal Development of Insects. London: Oliver and Boyd, 1965.
Dobzhansky, T.: Genetics of the Evolutionary Process. New York: Columbia University Press, 1970.
Farris, J.S.: Estimating phylogenetic trees from distance matrices. Amer. Natur. 106:645-667. (1972).
Finot, A.: Les Orthoptères de la France. Paris: F. Deyrolle, 1883.
Folsom, J.W., Woke, P.A.: The field cricket in relation to the cotton plant in Louisiana. USDA Tech. Bull. 642:1-28. (1939).
Fontana, P.G., Hogan, T.W.: Cytogenetic and hybridization studies of geographic populations of *Teleogryllus commodus* (Walker) and *T. oceanicus* (Le Guillou) (Orthoptera:Gryllidae). Aust. J. Zool. 17:13-35. (1969).
Fulton, B.B.: Speciation in the field cricket. Evolution 6:283-295. (1952).
Grellet, P.: Cinétique du developpement embryonnaire des Gryllides. Bull. Biol. France Belg. 95:615-643. (1961).
Harper, P.P., Pilon, J.G.: Annual patterns of emergence of some Quebec stoneflies (Insecta:Plecoptera). Can. J. Zool. 48:681-694. (1970).
Harrison, R.G.: Patterns of variation and genetic differentiation in closely related species: The field crickets of eastern North America. Ph.D. Thesis, Cornell University.(1977).
Hill, K.G., Loftus-Hills, J.J., Gartside, D.F.: Premating isolation between the Australian field crickets *Teleogryllus commodus* and *T. oceanicus* (Orthoptera:Gryllidae). Aust. J. Zool. 20: 153-163. (1972).
Hogan, T.W.: Some diapause characteristics and interfertility of three geographic populations of *Teleogryllus commodus* (Walker) (Orthoptera:Gryllidae). Aust. J. Zool. 13:455-459. (1965).
Hoy, R.R., Hahn, J., Paul, R.C.: Hybrid cricket auditory behavior: Evidence for genetic coupling in animal communication. Science 195:82-84. (1977).
Hynes, H.B.N.: Biology of Plecoptera. Ann. Rev. Ent. 21:135-154. (1976).
Knerer, G., Atwood, C.E.: Diprionid sawflies: Polymorphism and speciation. Science 179:1090-1099. (1973).
Lees, A.D.: The Physiology of Diapause in Arthropods. Cambridge: Cambridge University Press, 1955.

Leroy, Y.: Etude du chant de deux espèces de Grillons et de leur hybride (*Gryllus commodus* Walker, *Gryllus oceanicus* Le Guillou, Orthoptères). C.R. Acad. Sci. 256:268-270. (1963).
Leroy, Y.: Signaux acoustiques, comportement et systématique de quelques espèces de Gryllides (Orthoptères, Ensifères). D.Sc. Thesis, University of Paris. (1966).
Lim, H.-C., Vickery, V.R., Kevan, D.K.McE.: Cytological studies of Antipodean *Teleogryllus* species and their hybrids (Orthoptera: Gryllidae). Can. J. Zool. 47:189-196. (1969).
Lim, H.-C., Vickery, V.R., Kevan, D.K.McE.: Cytogenetic studies in relation to taxonomy within the family Gryllidae (Orthoptera). I. Subfamily Gryllinae. Can. J. Zool. 51:179-186. (1973).
MacFarlane, J.R., Drummond, F.H.: Embryonic diapause in a hybrid between two Australian species of field crickets, *Teleogryllus* (Orthoptera:Gryllidae). Aust. J. Zool. 18:265-272. (1970).
Masaki, S.: Geographic variation of diapause in insects. Bull. Fac. Agric. Hirosaki Univ. 7:66-98. (1961)
Masaki, S.: Geographic variation in the intrinsic incubation period: A physiological cline in the Emma field cricket (Orthoptera:Gryllidae:*Teleogryllus*). Bull. Fac. Agric. Hirosaki Univ. 11:59-90. (1965).
Masaki, S.: Geographic variation and climatic adaptation in a field cricket (Orthoptera:Gryllidae). Evolution 21:725-741. (1967).
Masaki, S., Ohmachi, F.: Divergence of photoperiodic response and hybrid development in *Teleogryllus* (Orthoptera:Gryllidae). Kontyû 35:83-105. (1967).
Mayr, E.: Animal Species and Evolution. Cambridge, Massachusetts: Belknap Press, 1963a.
Mayr, E.: Reply to criticism by R.D. Alexander. Syst. Zool. 12: 204-206. (1963b).
McGregor, E.A.: The true cricket - a serious cotton pest in California. USDA Circular No. 75:1-8. (1929).
Monteith, L.G.: Crickets as predators of the apple maggot *Rhagoletis pomonella* (Diptera:Tephritidae). Can. Ent. 103:52-58. (1971).
Nebeker, A.V.: Effect of temperature at different altitudes on the emergence of aquatic insects from a single stream. J. Kansas Ent. Soc. 44:26-35. (1971).
Nei, M.: Genetic distance between populations. Amer. Natur. 106: 283-292. (1972).
Ohmachi, F., Masaki, S.: Interspecific crossing and development of hybrids between the Japanese species of *Teleogryllus* (Orthoptera: Gryllidae). Evolution 18:405-416. (1964).
Ohmachi, F., Matsuura, H.: On the Japanese large field cricket and its allied species. Bull. Fac. Agric. Mie Univ. 2:63-72. (1951) (In Japanese).
Rakshpal, R.: Diapause in the eggs of *Gryllus pennsylvanicus* Burmeister (Orthoptera:Gryllidae). Can. J. Zool. 41:179-194. (1962).
Randell, R.L., Kevan, D.K.McE.: A cytological study of certain American species of *Gryllus* Linne (Orthoptera:Gryllidae) and their hybrids. Ann. Soc. Ent. Quebec 7:48-59. (1962).
Severin, H.C.: The common black field cricket - a serious pest in South Dakota. South Dakota Exp. Sta. Tech. Bull. 295:1-51. (1935).
Smith, D.S.: Notes on the destruction of grasshopper eggs by field cricket *Acheta assimilis luctuosus* (Serville) (Orthoptera: Gryllidae). Can. Ent. 91:127. (1959).
Sparks, A.N., Chiang, H.C., Keaster, A.J., Fairchild, M.L., Brindley, T.A.: Field studies of European corn borer biotypes in the midwest. J. Econ. Ent. 59:922-928. (1966).
Tauber, M.J., Tauber, C.A.: Environmental control of univoltinism and its evolution in an insect species. Can. J. Zool. 54:260-265.

(1976).

Tauber, C.A., Tauber, M.J., Nechols, J.R.: Two genes control seasonal isolation in sibling species. Science 197:592-593. (1977).

Thomas, N.A., Reed, L.B.: The field cricket as a pest of strawberries and its control. J. Econ. Ent. 30:137-140. (1937).

Walker, T.J.: *Gryllus ovisopis* n. sp.: An egg-diapausing univoltine cricket with no calling song (Orthoptera:Gryllidae). Fla. Ent. 56:13-22. (1974).

White, M.J.D. (ed.): Genetic Mechanisms of Speciation in Insects. Sydney: Australia and New Zealand Book Co., 1974.

Some Contributions of Snails to the Development of Ecological Genetics

B. CLARKE

A. Introduction

It has been suggested that the British school of ecological genetics sprang from an upper-middle-class fascination with snails and butterflies (Lewontin, 1974). My task, in the following account, is to argue that, for the molluscs at least, this fascination has a basis in scientific logic.

Studies on the ecological genetics of snails have recently been reviewed at length (Murray, 1975; Jones et al., 1977; Clarke et al., 1978), and it would not be profitable to do so again. These studies have, however, raised some issues that are relevant in a wider context. They are particularly relevant to the interpretation of enzyme polymorphisms, because in this endeavour also there has been a progress from an initial belief in the pervading importance of random genetic drift to a slow accumulation of evidence for natural selection. It seems that similar progress is about to occur in the study of nucleic acid sequences, and we may hope (albeit without too high a degree of expectation) that the lessons learned among the snails and butterflies may be applied among the molecules. The most important lesson is that we cannot hope to understand the causes of genetic variation without a knowledge of the ecology of the organisms concerned.

For the ecological geneticist, whatever may be his socio-economic status, the advantages of snails are manifold. They are highly polymorphic and polytypic, easy to collect and generally tolerant of laboratory conditions. They can be marked quickly and permanently. When they die they leave behind their shells as memorials of their genotypes, and sometimes of their modes of death. Their populations are small in area, but often not in numbers, and remain in one place over many generations. Most species are hermaphrodites, and some are viviparous. Indeed, their only serious disadvantage is a relatively long generation time (from three months upwards in the laboratory, and longer in the natural condition). It is not surprising that they have received attention.

The results of this attention have contributed to our understanding of many problems in ecological genetics. Of these I propose to consider four; first, the problem of how natural selection may be detected; second, the problem of how linkage disequilibrium may be generated; third, the problem of how polymorphisms may be maintained; and fourth, the problem of how species may have originated.

B. Detecting Natural Selection

The species that have been studied in the most detail are members of the genus *Cepaea*. Their very obvious colour and banding polymorphisms attracted early interest, and they were among the first organisms in which Mendel's rediscovered laws were confirmed (Lang, 1908). They were also among the first to have been studied systematically by ecological geneticists (Diver, 1932). The early workers almost unanimously believed that spatial variations in the frequencies of colour and banding genotypes were entirely due to random processes (Diver, 1939). This view reached its peak with the sophisticated studies by Lamotte (1951) who showed that the patterns of frequencies within and between populations could adequately be explained using Wright's theory of random genetic drift. Although Lamotte found it necessary to postulate rather high mutation rates (of the order of 10^{-4}), it seems clear that modern developments in the theory of neutral substitutions (and intragenic recombination) would no longer require him to do so. Despite this satisfactory agreement between theory and reality, Cain and Sheppard (1954), Lamotte himself (1959), and others have shown that selective forces are acting on the polymorphisms in populations whose frequency distributions agree with those derived from the 'neutral' model.

This nicely illustrates the hazard of supposing that because an observed distribution approximately corresponds to a theoretical one, the agreement between them necessarily supports the underlying theory. The most that can be said is that it does not refute it. It is possible that present-day advocates of neutrality in enzyme polymorphisms have not always appreciated the magnitude of this hazard.

The evidence for natural selection acting on the *Cepaea* polymorphisms is now well known, and very strong. It began to appear as soon as the habitats of the organisms were carefully observed. Cain and Sheppard (1950, 1954) found clear associations in *Cepaea nemoralis* between the frequencies of the colour and banding genotypes and the nature of the habitat (particularly the colour of the background). Brown and unbanded snails were relatively more common in woods, whereas yellow and unbanded snails were relatively more common in grasslands and 'rough herbage'. These associations were strong, and were found in many independent pairs of woods and grasslands. Cain and Sheppard interpreted them in terms of selective predation, particularly by thrushes, and argued that the predators tended to eat more of the snails whose colours and patterns stood out against the background. This interpretation was greatly strengthened by Sheppard's (1951) elegant experiments demonstrating that thrushes do indeed predate selectively on the colour morphs of *C. nemoralis* (see also Murray, 1962; Wolda, 1963; Arnold, 1966; Carter, 1968a; Cain and Currey, 1968). Correlations with habitat that can be interpreted in terms of selective predation have been found in several other species of snails including *Cepaea hortensis* (Clarke, 1960; Carter, 1968b), *Arianta arbustorum* (Parkin, 1971), and *Hygromia striolata* (Jones, Briscoe and Clarke, 1974). A very striking correlation has been reported in *Cochlicella acuta* by Lewis (1977). Banded morphs of this species are more frequent in the stabilized regions of sand dunes (on short turf) than in less stable (and sandier) parts. Lewis's observations give particularly strong evidence of natural selection because the association is found repeatedly over a large part of the species' range (covering a distance of more than 1000 miles). An interpretation in terms of visual selection, however, is more difficult than in the previously-mentioned species. At first sight, banded morphs of *C. acuta* seem to be more conspicuous than unbanded on short turf, but to be relatively less conspicuous in sandier areas. It seems that other selective factors may be at work.

In the search for alternative factors acting on shell-colour polymorphisms, an obvious candidate is direct selection by incident radiation. Jones (1973), and Heath (1975) have shown that darker morphs reach higher temperatures (by 1 or 2°C) than do paler morphs when they are exposed to direct sunlight. Unfortunately there are not yet any statistically satisfactory experiments showing that these temperature differences are significant contributors to differential survival, although Richardson (1974) has produced strong indirect evidence that pink and five-banded *Cepaea nemoralis* are more prone to death by 'overheating' in a sand-dune habitat.

Many surveyors of spatial differences in the genotype frequencies of *Cepaea nemoralis* and *C. hortensis* have interpreted them in terms of selection by climatic factors (for reviews see Jones et al., 1977; Clarke et al., 1978). There are, however, statistical difficulties. It has not yet been demonstrated beyond doubt that particular correlations between climatic variables and morph-frequencies are genuine in the sense that they are repeated in a sufficiently large number of *independent* replicates (statistically, a minimum of six is required). The problem of independence is acute because we know that the morph-frequencies of snail populations are spatially correlated. If a directional cline occurs (such clines are common in snails) and if there is a directional gradient in some climatic factor, the morph-frequencies will necessarily be correlated with the climatic variable unless the two directions are exactly at right angles to each other. When there are many morphs and many factors, some strong correlations are inevitable, even if the clines have arisen entirely non-selectively. This difficulty occurs not only with clines over short distances but also with the long latitudinal clines reported for shell colour morphs in *Cepaea* (Jones, 1973) and for enzyme polymorphisms in other organisms. The problems of disentangling spatial correlations are discussed by Dr. R. R. Sokal in his contribution to this volume.

There are other difficulties. A correlation with climate, even if genuine, may only be indirect. Climate is bound to affect the distributions of many other relevant factors, such as background colours, flora, competitors, parasites and predators. An illustrative example is the study by Cain (1968) in which he surveyed sand-dune populations of *Cepaea nemoralis*. He classified twenty-four separate dune systems into 'simple' and 'complex' topographies. Those in the 'complex' class, eight in number, all had populations of *nemoralis* containing brown shells. Of the twenty-two dune systems in the 'simple' category, only two contained populations with browns. The difference is statistically significant, and strongly suggestive of natural selection. Cain argues that the causative agent is climatic, because a complex topography would allow the accumulation of pockets of cold air, in which browns are presumed to be at a selective advantage because of their colour. While this explanation may well be the correct one, it is likely that 'simple' and 'complex' dunes differ in a large number of other ways.

It is worth pointing out that the differences between woods and grasslands described by Cain and Sheppard (1950, 1954) might be due in part to direct climatic selection. The relative preponderance of yellows in hedgerows and grasslands is exactly what would be expected if darker colours are at a disadvantage in places where the snails are exposed to sunlight. However, the evidence for selective predation is a great deal stronger than that for climatic selection. It is stronger because we have direct and rigorous experimental evidence that predators do indeed eat snails selectively. Furthermore, we can understand this selection in terms of known facts about the eyesight of the birds, the colours of the snails, and the nature of the backgrounds on which they live. We have a complete chain of causal explanation from the gene to the ecology. In the case of climatic selection the chain is not yet complete.

If we are to believe that selection acts directly on a locus, rather than on other loci in linkage disequilibrium with it, we are required to establish every link in the causal chain. In other words, we need a clear *mechanistic* connection via the selective agent between the variation in the gene product and the fitness of the individual. This requirement applies equally to enzyme polymorphisms, as has been pointed out elsewhere (Clarke, 1975a,b). It is not sufficient to carry out mathematical surveys of gene frequencies. We must study the polymorphisms themselves, and attempt to understand their individual mechanics.

C. Linkage Disequilibrium

Evidence of close linkage between different loci determining shell colour and pattern polymorphisms has been found in every species of snail for which the appropriate crosses have been made. It occurs in *Arianta arbustorum* (Cook and King, 1966), *Cepaea nemoralis* and *C. hortenis* (many authors, reviewed by Murray, 1975), *Cochlicella acuta* (Lewis, 1968), *Partula taeniata* (Murray and Clarke, 1976a), *Limicolaria flammulata* and *L. aurora* (Barker, 1968, 1969) and *Bradybaena similaris* (Komai and Emura, 1955). Widespread linkage disequilibrium is found within populations of all these species.

In *Cepaea nemoralis* the direction of disequilibrium between colour (pink vs. yellow) and banding (unbanded vs. banded) can depend upon the habitat. This happens in the region of Oxford, England, where woodland populations show excesses of pink unbanded and yellow banded shells, and deficiencies of pink banded and yellow unbanded. In hedges and grasslands the disequilibrium can go in either direction (Cain and Sheppard, 1954). These patterns are explicable in terms of visual selection because against a woodland background of brown leaf-litter the pink unbanded shells are the most cryptic of the four possible types, and the yellow unbanded the least so. For yellow snails in woods, banding greatly improves crypsis, but for pink snails it does not. Among populations in less visually uniform habitats, like hedgerows and grasslands, the directions of relative advantages and disadvantages are less predictable; and the disequilibria vary in direction.

Because this is one of the few cases where linkage disequilibrium can be attributed to known selective interactions between loci, it is remarkable that recent discussions of the phenomenon (for example by Lewontin, 1974) have ignored it. The case is even more remarkable when we remember that it was *Cepaea* that generated the first modern mathematical model of selective interaction between loci (Kimura, 1956). The explanation may lie in the intellectual isolation generated by the Atlantic Ocean, intensified by a difference in terminology. British workers have tended to describe complexes of loci in linkage disequilibrium as 'supergenes'. This is a convenient usage, but it may perhaps have caused the omission of some relevant papers from computer-based surveys of the literature. The same fate seems to have befallen the cases of linkage disequilibrium in the colour and pattern polymorphisms of locusts (Nabours, 1929; Nabours et al., 1933) and butterflies (Sheppard, 1959; Clarke and Sheppard, 1972). Such examples are particularly interesting at a time when it appears that, with respect to enzyme polymorphisms, linkage disequilibrium is not after all a common phenomenon (except when the enzyme loci are associated with inversions).

The relatively high frequency of disequilibria among loci for visible characters has two possible, non-exclusive, explanations. The first is that the selective interactions between 'visible' loci are on the average a great deal stronger than those between enzyme loci. The second is that colour polymorphisms in snails, and other organisms, may reflect the segregation of chromosomal inversions, or of other methods of crossover suppression. With respect to the second explanation it is perhaps significant that Johnson (1976) has reported a disequilibrium between an enzyme locus (*malate-dehydrogenase-1*) and the colour-banding supergene in *Cepaea nemoralis*.

The chromosomes of *C. nemoralis* have been regarded as too small and numerous (n = 22) for detailed study, and there has been little hope of detecting inversions and other chromosomal variants cytologically. Recently, however, as a result of work by Bantock and his colleagues (Bantock, 1972; Price, 1974, 1975a, 1975b; Price and Bantock, 1975) there have been great improvements in the techniques of staining and preparation. They have led to some tantalizing hints of associations between the extent of shell polymorphism and the frequency of chiasmata on the largest chromosome.

D. The maintenance of polymorphism

Because we know that at least some shell polymorphisms in snails are subject to directional selection, we are faced with the problem of how these polymorphisms are maintained within populations.

When the selection was first discovered, it was argued that the polymorphisms must be the result of heterozygous advantage (Cain and Sheppard, 1954). The case for this type of maintenance is still vigorously argued (Ford, 1975). However, because the genes for colour and pattern usually show complete dominance, it has been difficult to find evidence of it. The only case known to me is the report by Komai and Emura (1955) that double heterozygotes for colour and banding in *Bradybaena similaris* have faster growth rates than homozygotes. Both heterozygotes and homozygotes were the offspring of crosses between brown banded shells, but Komai and Emura did not record the origins of the parents. Thus it is possible that their observation was due to the chromosomes carrying the different alleles having originated in different populations. In other words, it might have been due to chromosomal heterosis rather than to heterozygous advantage at the colour and banding loci themselves.

Cain and Sheppard (1954) suggested, in relation to the polymorphism in *Cepaea nemoralis*, that predators might concentrate on common varieties to such an extent that rarer varieties would gain an advantage. At the time there seemed to be no experimental evidence that predators could indeed give rise to such frequency-dependent selection on polymorphic prey (except, necessarily, on polymorphic Batesian mimics), and Cain and Sheppard concluded that this kind of selection was unlikely to be important in maintaining the polymorphism. Their suggestion led, however, to a series of investigations in which it soon became apparent that predators do tend to concentrate disproportionately on common varieties and to ignore rare ones. This behaviour seems to occur regularly in fish (Clarke, 1962a; Maskell et al., 1977), birds (Allen and Clarke, 1968; Croze, 1970; Allen, 1972a, 1974, 1977; Manly, Miller and Cook, 1972; Cook and Miller, 1977) and small mammals (Soane and Clarke, 1973). It is found not only when predators are hunting different varieties of one species, but also when they are hunting different species (Tinbergen, 1960; Gibb, 1962). This phenomenon has been called 'apostatic selection' (Clarke, 1962a) and 'predator switching' (Murdoch, 1969). It was the first example of frequency-dependent selection to be investigated experimentally, and remains the best understood.

One bird that is known to predate in a frequency-dependent manner is the Song Thrush, a major predator of *Cepaea* and other polymorphic snails (Allen, 1972a; Harvey, Birley and Blackstock, 1975). Consequently there is strong reason to believe that apostatic selection plays a part in the maintenance of molluscan colour and pattern polymorphisms, particularly when we remember that this type of selection provides an explanation (which heterozygous advantage does not) for the *visual* distinctness of the morphs (Clarke, 1962a). However, most of the experiments on apostatic selection have used artifical or vegetable prey, and a formal demonstration that there is a frequency-dependent component in the predation of snails under natural conditions has not yet been published (but see Harvey, Birley and Blackstock, 1975).

Since apostatic selection by predators seems to be a very general phenomenon, it may well contribute to initiating and maintaining colour polymorphisms in many other animal groups and perhaps in some plants (Clarke, 1962a, 1969). Parasites probably act in a similar manner, becoming adapted to the commonest host genotype and disproportionately attacking it (Haldane, 1949). It has been suggested that frequency-dependent selection by parasites is a major factor in maintaining enzyme polymorphisms (Clarke, 1976).

We may note that selection by predators is, and selection by parasites is likely to be, density-dependent as well as frequency-dependent (Clarke, 1962a, 1975a,b; Allen, 1972b; Cook and Miller, 1977; Horsley, 1978). At very high densities it seems that rare colour varieties may actually stand out against the 'background' produced by the common ones, and consequently be over-predated (Allen, 1972b; Horsley, 1978). This generates selection for uniformity rather than polymorphism. At very low densities even the commonest form is not likely to be encountered often enough for the predator to develop a preference for it. Once more, uniformity is likely to result.

The behaviour of predators that maintains polymorphism within one species of prey should produce divergence between two species (as long as they share the same predator). In other words, apostatic selection should be a factor promoting character displacement. Some evidence of such displacement has been found in sympatric populations of *Cepaea nemoralis* and *Cepaea hortensis* (Clarke, 1962b). The evidence has been disputed (Carter, 1967) and defended (Clarke, 1969). Whatever the outcome of this particular debate, the expectation of displacement remains. Similarly, we may expect parasites to promote biochemical divergence between host species.

E. The origin of species

Various studies of *Cepaea nemoralis* have shown that the pattern of variation observed by Cain and Sheppard (1954) in the Oxford region, where genotype frequencies respond to habitat, is not universal. There are other regions in which *C. nemoralis* shows the preponderance of a few morphs over areas very much larger than the panmictic unit, apparently regardless of habitat or background. Between such areas the morph frequencies may change dramatically within distances of 200 metres or less, often in apparently uniform environments (Cain and Currey, 1963). These 'area effects' are not confined to *C. nemoralis*, but also occur in *C. hortensis* (Clarke, 1961), *Bradybaena similaris* (Komai and Emura, 1955), *Cerion* species (Mayr and Rosen, 1956), *Partula taeniata* (Clarke, 1968), and *Partula suturalis* (Clarke and Murray, 1971).

Cain and Currey (1963) have argued that the area effects in *C. nemoralis* may be due to cryptic environmental differences (probably climatic, see above), whose effects can produce sharp changes of selective forces over short distances. They believe that random genetic drift cannot be the present cause of the pattern because the areas are too large. Goodhart (1963) suggests that the differences between the areas are the consequences of the'founder principle', that the first groups of snails arriving in each area were small allopatric populations, differing by chance, and that initial differences between the populations persisted because of the evolution of balanced gene-complexes. Recent simulations by Endler (1977) indicate that patterns closely resembling area effects can develop purely as a result of random genetic drift in a series of small contiguous populations partially isolated from each other, although the differences in gene frequencies between areas are not as great as those observed in nature. Clarke (1966) has pointed out that the evolution of balanced gene-complexes does not require 'founder' populations. An algebraic model of a morph-ratio cline showed that area effects could evolve in a series of contiguous populations as a result of the accumulation of interacting modifier genes. There should then be an evolutionary trend towards the steepening of clines between them. Support for this model has come from work on experimental clines in *Drosophila* by Endler (1977), and on natural clines in *Partula* by Clarke and Murray (reported by Clarke, 1968). In both cases there was evidence that coadapted gene complexes had evolved at the two ends of the cline, but that the middle represented an area in which the coadaptations had broken down.

Further information has recently come from a study by Johnson (1976) on area effects in *Cepaea nemoralis* differing in the frequency of the single-banded (00300) gene. Johnson examined nine other polymorphic loci (six enzyme loci, two banding loci, and the colour locus). He found that three of these loci (*esterase-F*, *6-phosphogluconate-dehydrogenase* and *spread bands*) were associated with the area effects, indicating that the areas may differ in a significant proportion of their genes.

The steepening of a morph-ratio cline may continue to the point where the balanced gene complexes on either side of it are no longer genetically compatible. There would then be selection for reproductive isolation between them, and parapatric speciation would follow. Clarke and Murray (1969) have argued that the pattern of subspecies and species of *Partula* on the Pacific island of Moorea is explicable in these terms. Two forms may be clearly distinct and reproductively isolated in one region, but hybridize freely or intergrade in another (Murray and Clarke, 1968). Such changes can take place within 200 metres or less, and it is possible to find genetic connections, at one place or another (and sometimes through other forms) between eight of the eleven named species on the island. Recently it has been found that although the species differ very greatly in morphology and ecology, and although at any one place in Moorea as many as four of them may coexist sympatrically without hybridizing, nevertheless they seem to be indistinguishable at the level of conventional starch gel electrophoresis (Johnson, Clarke and Murray, 1977). This is not because they are lacking in heterozygosity, which in fact is unusually high (between 13.4 and 17.5 percent, averaging over twenty presumptive loci), but because their gene-frequencies are very similar. The close resemblance between *Partula suturalis* and *Partula taeniata* (with an average coefficient of genetic identity [Nei, 1972] of 0.92) is particularly remarkable because they coexist throughout the island without apparently interbreeding. They are also morphologically and ecologically very different, and they have different modes of expression and inheritance in their colour polymorphisms (Murray and Clarke, 1976a, 1976b).

Cepaea nemoralis provides a striking contrast. Its morphology is relatively uniform, but different populations seem to vary about as much in their enzymes as do different species of *Partula* (Brussard and McCracken, 1974; Brussard, 1975; Johnson, 1976).

These observations, like those of Turner (1974), King and Wilson (1975), and Johnson et al. (1975), indicate that the conditions for morphological divergence are not necessarily the same as the conditions for divergence at the protein level. It is tempting to suppose that the difference between them is due to the fact that protein variants reflect differences in structural genes, whereas morphological variants reflect, at least in part, differences in regulatory elements. However, before we can be sure that this is so we will need to know a great deal more, not only about genetic control in eukaryotes, but also about the ecological genetics of both kinds of variation.

F. Concluding Remarks

The study of snails has provided some of the best examples of selection acting in natural populations, and has added to our understanding of how such selection may be detected. It has also revealed several cases of strong linkage disequilibrium, and reasons for believing that they are due to selective interactions between loci. It has led to the demonstration of frequency-dependent selection by predators, and to a better perception of the mechanisms that maintain balanced polymorphisms. Finally, it has revealed some very curious patterns of spatial variation within and between species. An interest in snails has still, I believe, a part to play in the development of ecological genetics.

Because of the comment by Lewontin quoted at the beginning of this paper, I cannot resist taking, in the way of a conclusion, the dedication from Rimmer's (1907) *Land and Freshwater Shells of the British Isles*. "To those of my fellow-countrymen among the working classes who wisely employ their leisure hours in the pursuit of useful and elevating knowledge, with the hope that others, among their ranks, may be induced to forsake the paths of profitless and degrading dissipation, this volume is, with every good wish, dedicated by the author."

In the circumstances, I would like to extend this invitation beyond the limits of my native land.

Acknowledgements. I am very grateful to Dr. A. G. Clarke for critically reading the manuscript, to the Science Research Council for financial support, and to Dr. D. T. Parkin for the facilities that he has provided.

References

Allen, J.A.: Apostatic selection: The responses of wild passerines to artificial prey. Ph.D. Thesis, University of Edinburgh. (1972a).

Allen, J.A.: Evidence for stabilizing and apostatic selection by wild blackbirds. Nature 237:348-349. (1972b).

Allen, J.A.: Further evidence for apostatic selection by wild passerine birds: Training experiments. Heredity 33:361-372. (1974).

Allen, J.A.: Further evidence for apostatic selection by wild passerine birds, 9:1 experiments. Heredity 36:173-180. (1976).

Allen, J.A., Clarke, B.: Evidence for apostatic selection by wild passerines. Nature 220:501-502. (1968).

Arnold, R.W.: Factors affecting gene-frequencies in British and continental populations of *Cepaea*. Ph.D. Thesis, University of Oxford. (1966).

Bantock, C.R.: Localization of chiasmata in *Cepaea nemoralis* L. Heredity 29:213-221. (1972).

Barker, J.F.: Polymorphism in West African snails. Heredity 23: 81-98. (1968).

Barker, J.F.: Polymorphism in a West African snail. Amer. Natur. 103:259-266. (1969).

Brussard, P.F.: Geographic variation in North American colonies of *Cepaea nemoralis*. Evolution 29:402-410. (1975).

Brussard, P.F., McCracken, G.F.: Allozymic variation in a North American colony of *Cepaea nemoralis*. Heredity 33:98-101. (1974).

Cain, A.J.: Studies on *Cepaea*. V. Sand-dune populations of *Cepaea nemoralis* (L.). Phil. Trans. R. Soc. Ser. B 253:499-517. (1968).

Cain, A.J., Currey, J.D.: Area effects in *Cepaea*. Phil. Trans. R. Soc. Ser. B 246:1-81. (1963).

Cain, A.J., Currey, J.D.: Studies on *Cepaea*. III. Ecogenetics of a population of *Cepaea nemoralis* (L.) subject to strong area effects. Phil. Trans. R. Soc. Ser. B 253:447-482. (1968).

Cain, A.J., Sheppard, P.M.: Selection in the polymorphic land snail *Cepaea nemoralis*. Heredity 4:275-294. (1950).

Cain, A.J., Sheppard, P.M.: Natural selection in *Cepaea*. Genetics 39:89-116. (1954).

Carter, M.A.: Selection in mixed colonies of *Cepaea nemoralis* and *Cepaea hortensis*. Heredity 22:117-139. (1967).

Carter, M.A.: Thrush predation of an experimental population of the snail *Cepaea nemoralis* (L.). Proc. Linn. Soc. Lond. 179:241-249. (1968a).

Carter, M.A.: Studies on *Cepaea*. II. Area effects and visual selection in *Cepaea nemoralis* (L.) and *Cepaea hortensis*. Phil. Trans. R. Soc. Ser. B 253:397-446. (1968b).

Clarke, C.A., Sheppard, P.M.: Further studies on the genetics of the mimetic butterfly *Papilio memnon* L. Phil. Trans. R. Soc. Ser. B 263:35-70. (1972).

Clarke, B.: Divergent effects of natural selection on two closely-related polymorphic snails. Heredity 14:423-443. (1960).

Clarke, B.: Some factors affecting shell colour polymorphism in *Cepaea*. Ph.D. Thesis, University of Oxford. (1961).

Clarke, B.: Balanced polymorphism and the diversity of sympatric species. In Taxonomy and Geography. (ed. D. Nichols). Oxford: Systematic Association, 1962a.

Clarke, B.: Natural selection in mixed populations of two polymorphic snails. Heredity 17:319-345. (1962b).

Clarke, B.: The evolution of morph-ratio clines. Amer. Natur. 100: 389-402. (1966).

Clarke, B.: Balanced polymorphism and regional differentiation in land snails. In Evolution and Environment. (ed. E.T. Drake). New Haven: Yale University Press, 1968.

Clarke, B.: The evidence for apostatic selection. Heredity 24:347-352. (1969).

Clarke, B.: The contribution of ecological genetics to evolutionary theory: Detecting the direct effects of natural selection on particular polymorphic loci. Genetics 79:101-113. (1975a).

Clarke, B.: Frequency-dependent and density-dependent natural selection. In The Role of Natural Selection in Human Evolution. (ed. F.M. Salzano). Amsterdam: North-Holland Publ. Co., 1975b.

Clarke, B.: The ecological genetics of host parasite relationships. In Genetic Aspects of Host-Parasite Relationships. (ed. A.E.R. Taylor and R. Muller). Oxford: Blackwell Scientific Publications, 1976.

Clarke, B., Arthur, W., Horsley, D.T., Parkin, D.T.: Genetic variation and natural selection in pulmonate molluscs. In The Pulmonates, Vol. 2. (ed. J. Peake). New York: Academic Press, 1978.

Clarke, B., Murray, J.: Ecological genetics and speciation in land snails of the genus *Partula*. Biol. J. Linn. Soc. 1:31-42. (1969).

Clarke, B., Murray, J.: Polymorphism in a Polynesian land snail, *Partula suturalis vexillum*. In Ecological Genetics and Evolution. (ed. E.R. Creed). Oxford: Blackwell, 1971.

Cook, L.M., King, J.M.B.: Some data on the genetics of shell-character polymorphism in the snail *Arianta arbustorum*. Genetics 53:415-425. (1966).

Cook, L.M., Miller, P.: Density-dependent selection on polymorphic prey - some data. Amer. Natur. 111:594-598. (1977).

Croze, H.T.: Searching images in carrion crows. Z. Tierpsychol. 5: 1-85. (1970).

Diver, C.: Mollusc genetics. Proc. VI Int. Congr. Genet. (Ithaca) 2: 236-239. (1932).

Diver, C.: Aspects of the study of variation in snails. J. Conch. 21: 91-141. (1939).

Endler, J.A.: Geographic Variation, Speciation and Clines. Princeton: Princeton University Press, 1977.

Ford, E.B.: Ecological Genetics. London: Methuen, 1975.

Gibb, J.A.: L. Tinbergen's hypothesis on the role of specific search images. Ibis 104:106-111. (1962).

Goodhart, C.B.: "Area effects" and non-adaptive variation between populations of *Cepaea* (Mollusca). Heredity 18:459-465. (1963).

Haldane, J.B.S.: Disease and Evolution. La Ricerca Scientifica Suppl. 19:68-76. (1949).

Harvey, P.H., Birley, N., Blackstock, T.H.: The effect of experience on the selective behaviour of song thrushes feeding on artificial populations of *Cepaea* (Held.). Genetica 45:211-216. (1975).

Heath, D.J.: Colour, sunlight and internal temperatures in the landsnail *Cepaea nemoralis*. Oecologia 19:29-38. (1975).

Horsley, D.T.: The role of avian predators in maintaining colour polymorphism. Ph.D. Thesis, University of Nottingham. (1978).

Johnson, M.S.: Allozymes and area effects on the western Berkshire Downs. Heredity 36:105-121. (1976).

Johnson, M.S., Clarke, B., Murray, J.: Genetic variation and reproductive isolation in *Partula*. Evolution 31:116-126. (1977).

Johnson, W.E., Carson, H.L., Kaneshiro, K.Y., Steiner, W.W.M., Cooper, M.M.: Genetic variation in Hawaiian *Drosophila*. II. Allozymic differentiation in the *D. planitibia* subgroups. In Isozymes IV. (ed. C.L. Markert). New York: Academic Press, 1975.

Jones, J.S.: Ecological genetics and natural selection in molluscs. Science 182:546-552. (1973).

Jones, J.S., Briscoe, D.A., Clarke, B.: Natural selection on the polymorphic snail *Hygromia striolata*. Heredity 33:102-106. (1974).

Jones, J.S., Leith, B.H., Rawlings, P.: Polymorphism in *Cepaea* - a problem with too many solutions? Ann. Rev. Ecol. Syst. 8:109-143. (1977).

King, M-C., Wilson, A.C.: Evolution at two levels in humans and chimpanzees. Science 188:107-116. (1975).

Kimura, M.: A model of a genetic system which leads to closer linkage by natural selection. Evolution 10:278-287. (1956).

Komai, T., Emura, S.: A study of population genetics on the polymorphic land snail *Bradybaena similaris*. Evolution 9:400-418. (1955).

Lamotte, M.: Recherches sur la structure génétique des populations naturelles de *Cepaea nemoralis* (L.). Bull. Biol. Fr. Belg. Suppl. 35:1-239. (1951).

Lamotte, M.: Polymorphism of natural populations of *Cepaea nemoralis*. Cold Spring Harb. Symp. Quant. Biol. 24:65-84. (1959).

Lang, A.: Uber die Bastarde von *Helix hortensis* Mueller und *H. nemoralis* L. Festschr. Univ. Jena. 1908:1-120. (1908).

Lewis, G.: Polymorphism in the shell characters of certain helicid molluscs. Ph.D. Thesis, University of Oxford. (1968).

Lewis, G.: Polymorphism and selection in *Cochlicella acuta*. Phil. Trans. R. Soc. Ser. b 276:399-451. (1977).

Lewontin, R.C.: The Genetic Basis of Evolutionary Change. New York: Columbia University Press, 1974.

Manly, B.F.J., Miller, P., Cook, L.M.: Analysis of a selective predation experiment. Amer. Natur. 106:719-736. (1972).

Maskell, M., Parkin, D.T., Verspoor, E.: Apostatic selection by sticklebacks upon a dimorphic prey. Heredity 39:83-89. (1977).

Mayr, E., Rosen, C.E.: Geographic variation and hybridisation in populations of Bahama snails *(Cerion)*. Amer. Mus. Novit. 1806:1-48. (1956).

Murdoch, W.W.: Switching in general predators: Experiments on predator specificity and stability of prey populations. Ecol. Monogr. 39:335-354. (1969).

Murray, J.: Factors affecting gene frequencies in some populations of *Cepaea*. Ph.D. Thesis, University of Oxford. (1962).

Murray, J.: The genetics of the Mollusca. In Handbook of Genetics. (ed. J.C. King). New York: Academic Press, 1975.

Murray, J., Clarke, B.: Partial reproductive isolation in the genus *Partula* (Gastropoda) on Moorea. Evolution 22:684-698. (1968).

Murray, J., Clarke, B.: Supergenes in polymorphic land snails. I. *Partula taeniata*. Heredity 37:253-269. (1976a).

Murray, J., Clarke, B.: Supergenes in polymorphic land snails II. *Partula suturalis*. Heredity 37:271-282. (1976b).

Nabours, R.K.: The genetics of the Tettigidae (Grouse Locusts). Bibliograph. Genet. 5:27-104. (1929).

Nabours, R.K., Larson, I., Hertwig, N.: Inheritance of color patterns in the grouse locust *Acrydium arenosum* Burmeister (Tettigidae). Genetics 18:159-171. (1933).

Nei, M.: Genetic distance between populations. Amer. Natur. 106: 283-292. (1972).

Parkin, D.T.: Visual selection in the land snail *Arianta arbustorum*. Heredity 26:35-47. (1971).

Price, D.J.: Variation in chiasma frequency in *Cepaea nemoralis*. Heredity 32:211-218. (1974).

Price, D.J.: Chiasma frequency variation with altitude in *Cepaea hortensis* (Mull.). Heredity 35:221-229. (1975a).

Price, D.J.: Position and frequency distribution of chiasmata in *Cepaea nemoralis* (L.). Caryologia 28:261-268. (1975b).

Price, D.J., Bantock, C.R.: Marginal populations of *Cepaea nemoralis* (L.) on the Brendon Hills, England. II. Variation in chiasma frequency. Evolution 29:278-286. (1975).

Richardson, A.M.M.: Differential climatic selection in natural populations of land snail *Cepaea nemoralis*. Nature 247:572-573. (1974).

Rimmer, R.: Land and Freshwater Shells of the British Isles. Edinburgh: John Grant, 1907.

Sheppard, P.M.: Fluctuations in the selective value of certain phenotypes in the polymorphic land snail *C. nemoralis*. Heredity 5:125-134. (1951).

Sheppard, P.M.: The evolution of mimicry; a problem in ecology and genetics. Cold Spring Harbor Symp. Quant. Biol. 24:131-140. (1959).

Soane, I.D., Clarke, B.: Evidence for apostatic selection by predators using olfactory cues. Nature 241:62-64. (1973).

Tinbergen, L.: The natural control of insects in pinewoods. Factors influencing the intensity of predation by song birds. Arch. Neerl. Zool. 13:265-343. (1960).

Turner, B.J.: Genetic divergence of Death Valley pupfish species: Biochemical versus morphological evidence. Evolution 28:281-294, (1974).

Wolda, H.: Natural populations of the polymorphic land snail *Cepaea nemoralis* (L.). Arch. Neerl. Zool. 15:381-471. (1963).

5 Plants

Genetic Demography of Plant Populations

M.T. Clegg, A.L. Kahler, and R.W. Allard

A. Introduction

Darwin (1958) remarks in his autobiography that the idea of natural selection came to him while reading for amusement Malthus' essay on *Population*. Malthus' recognition that populations tend to grow exponentially has become so ingrained in contemporary thinking that the simple statement of population growth tends to have an axiomatic character. Likewise, Darwin's insight that environmentally imposed limitations on population growth imply differential reproduction is today beyond dispute. Indeed, the very mechanism of adaptive evolution arises from environmentally mediated and genotypically differential controls on population growth. Despite this fundamental connection between genetics and demography and despite the early pioneering work of Norton (1926), serious attempts to develop an integrated theory of genetic demography have occurred only within the past decade (e.g. Anderson and King, 1970; Charlesworth, 1970, 1974; Charlesworth and Giesel, 1972; Demetrius, 1975).

Experimental studies of the interrelationships between demographic and genetic parameters are also in early stages. Until quite recently studies of genetic demography were confined largely to human populations (see e.g. Cavalli-Sforza and Bodmer, 1971; Ward and Weiss, 1976) and to a lesser extent to small animals (Tamarin and Krebs, 1969; Gaines and Krebs, 1971; Tinkle and Selander, 1973). Many features of plant population biology are particularly well suited to the investigation of genetic and demographic problems. These advantages include sedentary habit allowing the identification and study of individual plants through time, well developed methods for aging individuals belonging to woody species, and the fact that direct estimates of individual fertility can be made by counting annual seed output. In addition, the ecological relationships of individual plants are more easily described and quantified than are those of mobile animal species. Thus, it is natural to seek a union between plant ecology and plant genetics through the medium of demography.

This review considers the current status of the study of plant genetic demography from two perspectives. First, the general question of age specific survival for various life history stages is considered; and second, the techniques and problems associated with the estimation of life history components of selection are discussed. The available evidence from experimental and natural populations of plants often indicates large selective differences among genotypes in survival. Fewer attempts have been made to estimate fertility components of selection, although present data show that fertility selection is at least as important as viability selection.

B. Seed Dormancy

A natural point to begin tracing dynamics of the life history of plant populations is with the seed pool. Figure 1 presents a schematic representation to illustrate the critical life history stages through which plants develop. Unlike most animals, plant propagules enter a resting stage which may be brief (e.g. *Acer rubrum*, or *Poa annua* in northern latitudes) or which may last for tens of years or longer.

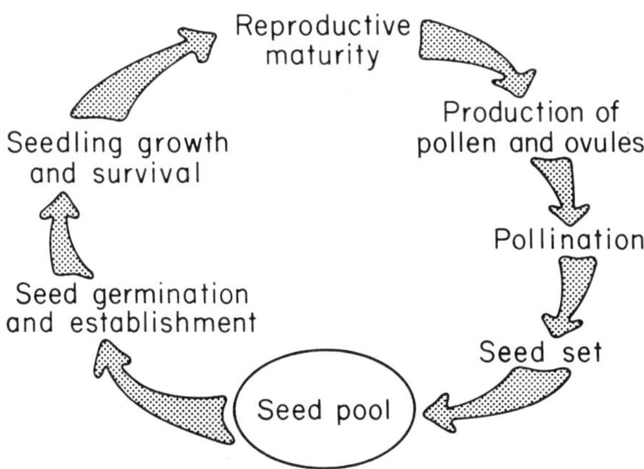

Fig. 1. Schematic representation of plant life history.

At first sight seed dormancy would appear to be maladaptive. This follows from the fact that seed dormancy, by delaying germination, delays the growth and reproduction of individuals and therefore reduces the population growth rate. These observations lead to the conclusion that a genotype with no dormancy, or reduced dormancy, would by virtue of its faster growth rate eventually replace genotypes with longer dormancy periods. Two lines of evidence suggest that populations often do not evolve to minimize dormancy. First, large differences in dormancy sometimes exist among plant populations within species (Jain, 1975); and second, genetic variation for seed dormancy can often be demonstrated within populations (e.g. Harper and McNaughton, 1960). The paradox of seed dormancy can be resolved by taking into account the complexities of the environment with its concomitant survival opportunities. Not all seasons of the year are opportune for plant growth and reproduction, and delaying germination may greatly enhance probabilities of survival and therefore enhance reproductive rates. Similarly, not all years are equally favorable; and a pool of dormant seed provides a buffer against exceptionally unfavorable years (Cohen, 1968; Levins, 1969).

Genetic variation in dormancy invests plant populations with a certain genetic inertia. The impact of viability and fertility selection on the genotypic frequency distribution among seed produced in any given year is modulated by the frequency distribution among seed carried over from previous years. If environmental fluctuations are large and erratic from year to year, the population will be buffered against a quick response to environmental noise. While several authors have commented on the potential role of seed dormancy in the maintenance of genetic variation (Jain, 1976; Harper and White, 1974; Antonovics, 1971; Epling, Lewis and Ball, 1960), this topic has not received detailed theoretical attention.

Studies of seed longevity in the soil (reviewed in Harper and White, 1974) suggest that survival probabilities are constant functions of time, leading to simple exponential decay curves. Unfortunately, little is known about genotype specific probabilities of survival in the seed pool or about genotype specific germination rates. Knowledge of both of these parameters is essential to a complete description of the genetic demography of plant populations.

C. Seed Germination and Establishment

Harper and his associates (Harper, 1965; Harper and White, 1970) have shown in an elegant series of experiments that seed germination is controlled by soil surface heterogeneity. According to these authors the soil surface is divided into a mosaic of fine-scale microhabitats, some of which are suitable for seed germination ("safe sites") and some of which are unsuitable ("unsafe sites"). The characteristics of microsites suitable for germination vary from species to species and may even vary according to seed morph within species. Thus, for example, plants of *Chenopodium album* produce four different seed morphs, each of which differs in germination requirements (Williams and Harper, 1965). However, the genetic basis of these morphological forms has not been established; and, in fact, the seed morph may be determined by the location of the seed on the maternal parent. The relationship between genetic diversity and microhabitat diversity in germination requirements remains a largely unexplored issue.

Following germination, plants enter a growth phase where both intra- and interspecific competition may be severe. Demographic data, in the form of life tables, provide a means of quantifying the impact of biotic and abiotic stress factors on population dynamics (Sharitz and McCormick, 1973). Furthermore, the demographic structure of a population may itself be channeled into one of several alternative adaptive modes. The theory of r and K selection (MacArthur, 1962) emphasizes alternative demographic strategies. According to this theory genotypes which allocate a large fraction of their resources to reproductive activities (r strategists) will be favored in environments characterized by high density independent mortality. Conversely, populations subjected to high density dependent mortality will be characterized by genotypes which devote a larger proportion of resources to vegetative rather than reproductive structures. These K selected populations will, therefore, be successful competitively, but at the expense of rapid population growth. Gadgil and Solbrig (1972), in an analysis of genetic variation in populations of dandelions (*Taraxacum officinale*), have shown that populations in highly disturbed environments feature genotypes with rapid growth rates and high reproductive outputs, whereas competitively superior genotypes, characterized by long vegetative periods, predominate in stable undisturbed habitats. These findings show that demographic parameters are adjusted by selection to increase the adaptation of a population to its environment.

D. Genotype Specific Age Structure

Techniques for aging many relatively long lived species are readily available. Consequently, age distributions can be determined in standing populations, which allows a partitioning of the genotypic frequency distribution into distinct age classes. Data of this kind can sometimes be used to make inferences about genetic changes during past time periods. For instance, Hiebert and Hamrick (personal communication) have partitioned genotypic frequencies by age class in bristle-cone pine (*Pinus aristata*). These trees are believed to be the oldest living organisms, with some individuals attaining ages of more than 4,000 years. In one area there appears to be evidence for the migration of populations from higher to lower elevations through time. Using age specific gene frequency data, it is possible to determine whether the lower elevation (younger) populations are representative of the higher elevation (source) populations, or whether migration and establishment occurred preferentially among some genotypes.

Schaal and Levin (1976), in one of the few plant population studies integrating genetics and demography, have investigated genetic change as a function of age structure. Genotypic frequencies were determined for 27 enzyme loci in the perennial herb *Liatris cylindracea*. Individual plants can be aged by counting rings of pigmented cells deposited annually in the corm. Plant ages ranged up to 44 years in the Illinois population selected for study. Six age classes were arbitrarily defined, and age specific genotypic frequencies were determined. Table 1 illustrates changes in genotypic numbers with age for the malic dehydrogenase locus in this study. The first concern is to establish whether the distribution of genotypic frequencies is independent of age class. This hypothesis is easily tested by means of a chi square contingency table. The test yields a chi square statistic of 107.95 with 10 degrees of freedom ($p < .001$). We, therefore, conclude that the genotypic frequency distribution is not the same among age classes. Much of the change in genotypic frequencies occurs among heterozygotes, which tend to increase in frequency with age. This latter fact is brought out by the F statistics that express observed heterozygosity relative to expected random mating heterozygosity.

The F statistics can also be viewed retrospectively. We can ask how the initial genotypic frequency distribution arose from the standing population. This question requires knowledge of the mating process and knowledge of genotypic frequencies among newly formed zygotes. Neither of these data is available; however, the large increase in F from the older to the younger ages suggests that inbreeding, due either to non-random union of gametes or to very small population size, is important. The average value of F, weighted by the frequency of each age class, is 0.486. Suppose that the value of F in age class 1 equals the value among newly formed zygotes. Further, suppose that each class is equally fertile and that mating occurs by a mixture of random outcrossing and self-fertilization (with probability s). Then the recurrence relation for the inbreeding coefficient, F, is

$$\underline{F}_t = \frac{s}{2}(1 + \underline{F}_{t-1}). \tag{1}$$

Assuming inbreeding is the only factor responsible for the high values of F observed in the youngest age class, we can substitute 0.486 for $\underline{F}_{t-1}$ and 0.625 for $\underline{F}_t$ and solve for s, which yields s = 0.84. Such a high value of s seems unreasonable on two grounds; first *L. cylindracea* is reported to be self-incompatable; and second, heterozygotes are reported to have a much higher reproductive potential (Schaal and Levin, 1976).

Table 1. Genotypic numbers according to age class computed from Schaal (1974), for a malic dehydrogenase locus (MDH-2). Gene frequencies, p = (1-q), and Wright's fixation index, $F = 1 - \frac{H}{2pq}$ where H denotes the relative frequency of observed heterozygotes, are reported in the last two rows.

	Age Class					
Genotype	1	2	3	4	5	6
$A_1 A_1$	94	253	478	231	110	29
$A_1 A_2$	18	92	71	98	64	24
$A_2 A_2$	22	89	62	83	41	15
Gene Frequency	0.77	0.69	0.84	0.68	0.66	0.61
$\underline{F}$	0.625	0.503	0.570	0.455	0.340	0.265

Can the extant frequency distribution be explained? The high values of F can be explained partially by the fact that gene frequencies are quite heterogeneous spatially in this population (Schaal, 1975). Spatial heterogeneity, with mating occurring largely among neighbors, causes the population to be divided into a number of small breeding units and thus contributes to homozygosity. Taking the F data at face value, though, it is necessary to suppose that the breeding units include two or fewer individuals. This can be seen by recalling that the expected increase in $\underline{F}$ in finite monoecious populations is

$$\underline{F}_t = \frac{1}{2N} + (1 - \frac{1}{2N})\underline{F}_{t-1}. \tag{2}$$

Solving for N using $\underline{F}_t = 0.625$ and $\underline{F}_{t-1} = 0.486$ yields N = 1.84. A complete explanation for the genotypic frequency behavior in this population clearly requires knowledge of genotypic frequencies among newly formed zygotes and knowledge of the mating cycle, a point which will be developed further in the discussion of fertility selection.

E. Viability Components of Selection

Demographic processes are most readily described in annual plant species. It is relatively easy to follow single cohorts through time in annual plants; and reproduction occurs as a more or less discrete episode followed by senescence. These features make annual plants ideal model systems for the study of the elementary processes of genetic demography. Clegg and Allard (1973), in a study of a natural population of the slender wild oat *(Avena barbata)*, found changes in genotypic frequency distributions between seedling and adult stages. These changes reflect the action of viability selection.

The viability component of selection is easily quantified by determining the set of scalars (relative viability) which map the seedling distribution into the adult distribution. Employing the notation established in Table 2, in which relative genotypic frequencies in

Table 2. Census points and notation for experimental design to measure viability and fertility components of selection in monoecious annual plant populations. The quantities f_{ij} and a_{ij} denote relative genotypic frequencies among the population at zygotic and adult stages. The numbers of progeny of genotype kl observed for the ij^{th} maternal genotype are denoted n_{kl}^{ij} where i, j, k, l = 1,2,...,n.

	Census Stage				
Genotype	Zygotic	Adult	A_iA_i	A_iA_j	Zygotic
A_iA_i	f_{ii}	a_{ii}	n_{ii}^{ii}	n_{ij}^{ii}	f'_{ii}
A_iA_j	$2f_{ij}$	$2a_{ij}$	n_{ii}^{ij}	n_{ij}^{ij}	$2f'_{ij}$
Sample Size	N_f	N_a			N'_f

adult and zygotic stages are denoted by a_{ij} and f_{ij} respectively, the a_{ij} are connected to the f_{ij} by the relation, $a_{ij} = v_{ij}f_{ij}$ (i,j=1,2,...,n). Hence, an estimator for the viability of the ij^{th} genotype is simply,

$$v_{ij} = a_{ij}/f_{ij}, \tag{3}$$

relative to $\sum_{i,j} v_{ij}f_{ij} = 1$. It is obvious from the expression for the relative viability estimator that selection estimates are ratios of random variables. The approximate variance of v_{ij} in large samples is,

$$\mathrm{Var}(v_{ij}) = \frac{a_{ij}}{f_{ij}^2}\left[\frac{(1-a_{ij})}{N_a} + \frac{a_{ij}(1-f_{ij})}{f_{ij}N_f}\right]. \tag{4}$$

This will be a large quantity, even in large samples. Table 3 illustrates this point with viability estimates and standard errors for three esterase loci in *A. barbata*.

Table 3. Viabilities relative to the mean viability ($\bar{V} = \sum_{ij} v_{ij} f_{ij} = 1$) for three esterase loci in *Avena barbata* (computed from Clegg and Allard, 1973). Standard errors are in parentheses.

	Viabilities				
Locus	V_{11}	V_{12}	V_{22}	N_f	N_a
E_4	1.08 (0.21)	1.15 (0.72)	0.95 (0.11)	66	85
E_9	0.79 (0.26)	2.05 (2.63)	0.99 (0.09)	66	101
E_{10}	1.09 (0.36)	1.03 (1.53)	0.95 (0.08)	66	101

In a study to determine whether there are differences among genotypes in viability and fertility components of selection, Clegg, Kahler and Allard (1977) obtained genotypic frequencies in large samples drawn from an experimental population of barley (more than 32,000 plants were assayed at four esterase loci). The population studied, barley Composite Cross V (CCV), was synthesized in 1937 by intercrossing 30 varieties of barley representing the major barley growing regions of the world. The 30 original parental varieties were crossed in pairs, and during the next three years the F_1 hybrids of each cycle were again pair crossed to produce a single grand hybrid stock involving all 30 parents. The F_2 generation of this grand hybrid population was grown in 1941. In every year since 1941, the population has been advanced following normal agricultural practices and without conscious selection. Seed from earlier generations have been saved by storing part of the seed harvest of each generation. Thus, barley CCV provides an exceptional opportunity for the experimental study of the genetics of plant populations.

Samples of seed and of adult plants with half-sib families were drawn from CCV following the design outlined in Table 2. Genotypic frequencies observed at seed and adult stages for the triallelic EC locus for two generations are shown in Table 4. The vectors of genotypic frequencies show large and significant differences between seed and adult stages, as judged by likelihood ratio tests. Viabilities, also shown in Table 4, are computed relative to $v_{11} = 1$ to compare estimates from different generations with different genotypic frequencies. The viability estimates quantify the extent to which the C_2C_2 and C_3C_3 genotypes are favored over the C_1C_1 genotype in survival from seed to adult stages. Clearly, there has been very strong selection between seed and adult stages resulting in a marked change in the distribution of genotypes entering the mating cycle.

Table 4. Genotypic frequencies, relative viabilities (standard errors in parentheses), and chi square statistics computed from likelihood ratio tests for generations 19 and 28 at locus EC in barley CCV (from Clegg, Kahler and Allard, 1977).

	Generation					
	19			28		
Genotype	f_{ij}	a_{ij}	v_{ij}	f_{ij}	a_{ij}	v_{ij}
C_1C_1	0.661	0.533	1.00	0.519	0.476	1.00
C_1C_2	0.0	0.001	--	0.009	0.003	0.46
						(0.27)
C_1C_3	0.002	0.005	3.43	0.005	0.004	0.76
			(2.80)			(0.49)
C_2C_2	0.058	0.096	2.07	0.205	0.256	1.36
			(0.35)			(0.14)
C_2C_3	0.002	0.0	0.0	0.002	0.0	0.0
C_3C_3	0.277	0.364	1.64	0.259	0.260	1.09
			(0.14)			(0.10)
Sample size	1007	1098		1102	1052	
$\chi^2(5)$		43.76			13.50	

F. Fertility Selection

The final demographic process to consider is fertility. Numerous investigations of the quantitative genetics of natural plant populations have consistently shown a large genetic variance for characters highly correlated with fertility (Allard, Jain and Workman, 1968; Hamrick and Allard, 1975). However, the construction of genotype-specific fertility schedules has not been attempted in plants for several reasons. A major difficulty associated with constructing genotype-specific fertility schedules arises when we attempt to partition fertility schedules by sex, because offspring produced depend upon mating pairs. In human demography this difficulty can be circumvented by considering offspring per female alone. Unfortunately, the device of attributing fertility to a single sex is not satisfactory for genetic demography because the process of gamete production and mating have a crucial influence on the way in which genotypes (the unit of concern in genetic demography) arise. Thus, as indicated by Figure 1, we must account simultaneously for the processes of mating, gamete production and gametic selection in describing the reproductive process. One way to approach this problem is to impose models of the reproductive system on the data and to estimate the fertility and mating parameters from those models that best fit the data.

Estimates of parameters relevant to genetic changes during the reproductive phases of the life cycle can be obtained from studies of seed produced by individual adult plants, following the design presented in Table 2 [see also Christiansen and Frydenberg (1973, 1976) for the use of mother-offspring sampling designs in detecting components of selection in mammal populations]. The sampling of seed parents and their progeny is easy in many plant species in which seeds remain attached to their maternal parent for a period of time following maturation. In such species data concerning the family structure of the population can be obtained; and these data provide information on the mating system of the population and partial information on selection during gametophytic stages of the life cycle. It is necessary to assume a mating structure for the population in order to make use of family data.

The mixed mating model is one that describes the reproductive biology of monoecious, self-compatible, plant species. This model assumes: (a) that both pollen and ovule in any mating originated from the same individual (with probability s), or (b) that the pollen in a mating come from the entire population of pollen (with probability t(= 1-s)) produced by all adults. Thus mating is divided between self-pollination and outcrossing. The second step in formalizing the mixed mating model requires an assumption regarding the probability distribution of pollen types in the pool from which outcross events are drawn. The most elementary assumption is that the probability, p_1, of drawing a pollen type, say A_1, is uniform over all maternal types in the population. With these assumptions it is a straightforward matter, given the maternal type, to deduce the conditional probabilities of obtaining the various progeny types. Table 5 presents these probabilities for an arbitrary number of alleles per locus.

Table 5. Probabilities under the mixed mating model of observing various progeny genotypes, given the maternal genotype and assuming an arbitrary number of alleles per locus ($i \neq j \neq k \neq \ell$, $i, j, k, \ell = 1,2,\ldots,n$).

Maternal Genotype	Progeny Genotype A_iA_i	A_iA_j	A_jA_j	A_iA_k	A_kA_l
A_iA_i	$s + t\, p_i$	$t\, p_j$	0	$t\, p_k$	0
A_iA_j	$\frac{s}{4} + \frac{t}{2} p_i$	$\frac{s}{2} + \frac{t}{2} (p_i + p_j)$	$\frac{s}{4} + \frac{t}{2} p_j$	$\frac{t}{2} p_k$	0

A noteworthy feature of the conditional probabilities displayed in Table 5 is that, given a heterozygous maternal parent, the probability of obtaining a heterozygous progeny is 1/2 for diallelic loci, independent of the mating system and independent of the distribution of pollen types. This fact provides a means of testing for ovule selection in the pool of gametes produced by heterozygotes, unconfounded by either pollen selection or mating system. Unfortunately, this property is a consequence of the symmetry of diallelic systems and does not extend to cases where there are more than two alleles at a locus.

The sampling design shown in Table 2 permits the estimation of the parameters entering the mating system model. Maximum likelihood estimation of t and pollen pool frequencies can be carried out using standard numerical procedures (Kahler, Clegg and Allard, 1975). Table 6 gives mating system estimates for family data sampled from two generations of the experimental barley population CCV. The estimates of t are small, consistent with the fact that barley is a predominantly self-fertilizing species. The estimates of pollen pool frequencies reflect the frequency of the pollen types involved in fertilization events (effective pollen pool). Two questions about pollen pool frequencies which arise immediately are: (1) is the assumption of a uniform distribution of pollen types reasonable? And (2) does the frequency of pollen types in the effective pollen pool equal the gene frequency among plants contributing pollen to the pool, i.e., has differential contribution of pollen by genotype or differential survival and functioning of pollen types altered the gene frequency distribution?

Table 6. Outcrossing rate (t), gene frequencies in the effective pollen pool (denoted p_{im}), among the adult population (denoted p_i) and in the pool of ovules (denoted P_{if}) for locus EC, in generations 19 and 28 in barley CCV. Standard errors are reported in parentheses. Data taken from Clegg, Kahler and Allard (1977).

		Gene Frequency					
Generation	t	p_1	p_{1m}	p_{1f}	p_2	p_{2m}	p_{2f}
19	0.011 (0.001)	0.54 (0.02)	0.49 (0.07)	0.66 (0.01)	0.10 (0.01)	0.09 (0.03)	0.06 (0.01)
28	0.011 (0.001)	0.48 (0.02)	0.41 (0.08)	0.54 (0.01)	0.26 (0.01)	0.10 (0.04)	0.22 (0.01)

The assumption of a uniform distribution of pollen types across the maternal population is most probably satisfied by barley CCV. This is because seeds were sown randomly in each year; and there is consequently little, if any, opportunity for spatial differentiation by genotype to occur. In natural populations, however, spatially determined gene frequency differentiation may seriously influence pollen pool estimates. In answer to the second question, Table 6 presents a sample of effective pollen pool and adult gene frequency estimates taken from barley CCV. Most contrasts between effective pollen pool and adult gene frequencies are nonsignificant (the large variances attached to pollen estimates permit the detection of only large differences). However, one contrast is highly significant (p_2 allele generation 28), indicating that the genetic composition of the effective pollen pool differs from that of the adult population. This difference might result either from differential pollen fertility among adult genotypes, or from selection among male gametophytes. Gametophytic selection, particularly among male gametophytes, is a common and well documented phenomenon in plant populations [see Mulcahy (1975) for a review of gametophytic selection]. In the present instance, however, the experimental design does not allow discrimination between these two components of selection.

To investigate the way selection affects the pool of ovules involved in fertilization events, we consider the transition equations that describe the mating process under the mixed mating model. The equations are:

$$f'_{ii} = s(u_{ii} + \frac{1}{2} \sum_{j \neq 1} u_{ij}) + t(p_{if}p_{im}), \quad i = 1,2,\ldots,n \qquad (5)$$

$$2f'_{ij} = s\, u_{ij} + t(p_{if}p_{jm} + p_{jf}p_{im}), \quad i \neq j,\ i,j = 1,2,\ldots,n \qquad (6)$$

where p_{if} and p_{im} denote gene frequencies among ovules and pollen, respectively, at the time of fertilization. The quantities u_{ij} are the relative frequencies among adults expected if all adults contribute ovules in equal proportion to the next generation. The f'_{ij} are the relative frequencies among newly formed zygotes. The quantities p_{im} are estimable from the mating system analysis. Similarly, gene frequencies among newly formed zygotes are $p'_i = \sum_{j=1}^{n} f'_{ij}$ and the gene frequencies among ovules involved in mating events are $p_{if} = \sum_{j=1}^{n} u_{ij}$. Thus an expression for p_{if} in terms of the observable quantities p_i, p_{im} and t can be obtained from equations (5) and (6) as,

$$p_{if} = \frac{2p'_i - t\, p_{im}}{2 - t}. \qquad (7)$$

Two assumptions require justification in imposing equation (7) on population data. First, the expression $p_{if} = \sum_j u_{ij}$ assumes that ovule production is the only factor that limits the fertility of self-pollinated plants (i.e. that pollen is abundant). Second, the estimate of p_i must come from a random sample of the newly formed population of zygotes. Subject to these two assumptions, relative frequencies of ovule types among the pool of uniting gametes can be estimated.

In barley CCV bulk seed were obtained from the harvest of each generation. The seeds were germinated in the laboratory under optimal conditions and assayed by electrophoresis seven days after germination. More than 99% of seeds produced assayable seedlings, so estimated genotypic frequencies closely reflect zygotic frequencies. The second assumption, that ovule production limits the fertility of selfed genotypes, is also in accord with barley reproductive biology. Estimates of gene frequencies in the pool of ovules, reported in Table 6, show that fertility selection has a large and important influence on gene frequency distributions. In many instances the gene frequencies among ovules differ from those estimated from the adult population contributing gametes to the next generation. Furthermore, the estimates of p_{if} and p_{im} often differ significantly (e.g. p_{im} versus p_{if} generation 19), indicating that the forces of selection affect the pool of ovules differently from the pool of pollen involved in outcross events.

We can also compute the hypothetical genotypic frequencies (u_{ij}) and compare these frequencies to the observed adult frequencies (a_{ij})

to determine the genotypic distribution of ovule fertilities. The functions which estimate the u_{ij} are:

$$u_{ii} = \frac{2}{s}\left[f'_{ii} - \left(\frac{2p_i - t\,p_{im}}{2 - t}\right)(t\,p_{im} + \frac{s}{2})\right], \quad i = 1,2,\ldots,n \qquad (8)$$

$$2u_{ij} = \frac{2}{s}\left\{2f'_{ij} - \frac{t}{2-t}\left[p_{jm}(2p_i - t\,p_{im}) + p_{im}(2p_j - t\,p_{jm})\right]\right\}$$

$$i \neq j$$

$$i,j = 1,2,\ldots,n \qquad (9)$$

Table 7 compares the observed adult frequencies (a_{ij}) with the hypothetical frequencies (u_{ij}) required to account for the mating

Table 7. The observed adult genotypic frequency distribution (a_{ij}), the hypothetical distribution which would exist if all adults were weighted by their ovule fertilities (u_{ij}) and fertility estimates (ℓ_{ij}). Standard errors are in parentheses. Data are for the EC locus in generations 19 and 28 of barley CVV (computed from Clegg, Kahler and Allard, 1977).

	Generation					
	19			28		
Genotype	a_{ij}	u_{ij}	ℓ_{ij}	a_{ij}	u_{ij}	ℓ_{ij}
C_1C_1	0.533	0.647	1.00	0.476	0.533	1.00
C_1C_2	0.001	0.001	0.72	0.003	0.0002	0.07
			(1.11)			(0.22)
C_1C_3	0.005	0.011	2.33	0.004	0.012	4.00
			(1.07)			(2.08)
C_2C_2	0.096	0.057	0.49	0.256	0.214	0.74
			(0.08)			(0.08)
C_2C_3	0.0	0.001	---	0.0	0.0	---
C_3C_3	0.364	0.282	0.62	0.260	0.243	0.80
			(0.06)			(0.08)
Sample size	1098	1078		1052	1139	

transition, assuming all maternal genotypes are equally fertile. The differences between the vectors a_{ij} and u_{ij} are highly significant for generation 19 (χ^2_5 = 19.9; $p < .005$), and border on significance for generation 28 ($\chi^2_4 = 9.2$; $p \approx .06$). (A caveat regarding the statistical tests must be raised in this case; the u_{ij} are functions of t and p_{im}, both of which are estimated and therefore subject to sampling error. Hence the linear transformation (8) does not preserve the probability structure of the f_{ij}. Although it is safe to regard the generation 19 frequencies as significantly different due to the very low probability, such a conclusion is unwarranted for the generation 28 data.) Clearly, the fertility component of selection must be taken into account in describing the genetic demography of plant populations.

To quantify the genotypic specific fertilities we denote the expected ovule production by ℓ_{ij} for the ij^{th} maternal genotype; then $u_{ij} = \ell_{ij}a_{ij}$, i.e., ℓ_{ij} is the scalar factor which carries the observed adult genotypic frequency distribution into the distribution that would exist if all adults contributed equally to the pool of ovules involved in fertilization events. Table 7 gives estimates of the u_{ij} and ℓ_{ij}, expressed relative to $\ell_{11} = 1$. The data show that fertility differences are sometimes large among the three homozygous genotypes at the triallelic EC locus (e.g., C_2C_2 generation 19).

It is also instructive to contrast the entries in Table 4 with those in Table 7. Comparison of the relative frequency vectors (f_{ij}), (a_{ij}) and (u_{ij}) reveals that genotypes which increase in frequency during the seedling to adult transition, presumably due to superior competitive ability, decline during reproductive phases of the life cycle. Thus a negative correlation exists between survival and reproductive success.

G. Questions of Interpretation

All the data on life history components of selection considered indicate strong selection at the marker loci under observation. Can we infer that gene substitutions at these marker loci have large effects on fitness during growth and reproduction? To answer this question, it is necessary to recall that each of the populations considered was heavily inbred. Inbreeding restricts genetic recombination and therefore allows the maintenance of strongly correlated blocks of genes (Allard et al., 1972; Allard, 1975; Clegg, Allard and Kahler, 1972). Consequently, the impact of selection will be transmitted throughout this correlational structure, and selection measured at any marker locus will reflect the cumulative effect of selection over many loci. The data show that strong selection acts on the variety of correlational structures segregating in the population. The data also show that this selection is manifest at each of the life history stages investigated.

In considering reproductive phases of the life cycle, we have imposed models on the data which purport to describe the processes of gamete production and mating. In choosing models to apply to the data, we are confronted by the conflicting demands of simplicity versus realism. Complicated models usually require extensive manipulation of experimental populations. Consequently, the gain in realism attendant upon a complex model is often more than offset by the artificial and unrealistic experimental designs required. Conversely, a very simple model may require unwarranted assumptions. The experimental population biologist must attempt to make the best compromise, based upon knowledge of the experimental materials and upon the objectives of the experiment.

In the present instance simple models of the reproductive system have been used to highlight and quantify the population components of mating and fertility. These models have the virtue that they exploit the information latent in the family structure of the population without requiring extensive manipulation of the population. They can easily be applied to natural as well as experimental populations. Barley CCV comes close to satisfying both the assumptions of the mixed mating model and the assumption that ovule production limits the fertility of selfed genotypes. Other plant populations may depart from this ideal to a greater or lesser degree. Nothing prevents the application of simple estimation models to situations where the validity of the assumptions are uncertain; however, interpretations of the parameter estimates must be made in the light of this uncertainty.

H. Conclusions

Unfortunately, only limited data are presently available on plant genetic demography. The paucity of data is particularly extreme when we consider reproductive phases of the life cycle. Although it is perhaps premature to generalize on such limited data, several points require comment. First, the available data reveal large viability and fertility components of selection in annual plant populations. The fertility component of selection appears to be as important as the viability component in determining total fitness. This finding is consistent with recent work with *Drosophila* which has provided evidence pointing to the importance of fertility selection and mating success (Bundgaard and Christiansen, 1972; Prout, 1971; Anderson and Watanabe, 1974).

A second noteworthy point is the finding that different genotypes appear to be favored at different stages of the life cycle. An implication of this finding is that studies which measure net selection over the entire life cycle will perforce underestimate the fraction of the population's reproductive potential which is absorbed in differential survival and differential fertility. Williams (1975), in attempting to account for the evolution of genetic recombination, has argued that differential selection is intense over various life cycle stages. The only way to investigate this claim is by studying selection as a demographic process.

The available data clearly show that the genetic demography of natural selection is not simple, even for annual plants. Any understanding of genetically mediated fitness differences must recognize that fitness is a compound parameter, arranged over the diverse life history phases of plants, and dependent on the specific environments experienced at each developmental stage. It is only through such a recognition that we can understand fitness as a functional and ecological problem.

Acknowledgements. We thank Dr. W. W. Anderson for comments on an earlier version of the manuscript.

References

Allard, R.W.: The mating system and microevolution. Genetics 79: 115-126. (1975).
Allard, R.W., Babbel, G.R., Clegg, M.T., Kahler, A.L.: Evidence for coadaptation in *Avena barbata*. Proc. Nat. Acad. Sci. U.S. 69: 3043-3048. (1972).
Allard, R.W., Jain, S.K., Workman, P.L.: The genetics of inbreeding populations. Advan. Genet. 14:55-131. (1968).
Anderson, W.W., King, C.E.: Age-specific selection. Proc. Nat. Acad. Sci. U.S. 66:780-786. (1970).
Anderson, W.W., Watanabe, T.K.: Selection by fertility in *Drosophila pseudoobscura*. Genetics 77:559-564. (1974).
Antonovics, J.: Population dynamics of the grass *Anthoxanthum odoratum* on a zinc mine. J. Ecol. 60:351-365. (1972).
Bundgaard, J., Christiansen, F.B.: Dynamics of polymorphisms. Genetics 71:439-460. (1972).
Cavalli-Sforza, L.L., Bodmer, W.F.: The Genetics of Human Populations. San Francisco: Freeman, 1971.
Charlesworth, B.: Selection in populations with overlapping generations. I. The use of malthusian parameters in population genetics. Theoret. Popul. Biol. 1:352-370. (1970).
Charlesworth, B.: Selection in populations with overlapping generations. VI. Rates of change of gene frequency and population growth rate. Theoret. Popul. Biol. 6:108-133. (1974).
Charlesworth, B., Giesel, J.T.: Selection in populations with overlapping generations. II. The relations between gene frequency and demographic variables. Amer. Natur. 106:388-401. (1972).
Christiansen, F.B., Frydenberg, O.: Selection component analysis of natural polymorphisms using population samples including mother-offspring combinations. Theoret. Popul. Biol. 4:425-445. (1973).
Christiansen, F.B., Frydenberg, O.: Selection component analysis of natural polymorphisms using mother-offspring samples of successive cohorts. In Population Genetics and Ecology. (ed. S. Karlin and E. Nevo). New York: Academic Press, 1976, pp. 277-301.
Clegg, M.T., Allard, R.W.: Viability versus fecundity selection in the Slender Wild Oat, *Avena barbata* L. Science 181:667-668. (1973).
Clegg, M.T., Allard, R.W., Kahler, A.L.: Is the gene the unit of selection? Evidence from two experimental plant populations. Proc. Nat. Acad. Sci. U.S. 69:2474-2478. (1972).
Clegg, M.T., Kahler, A.L., Allard, R.W.: Estimation of life cycle components of selection in an experimental plant population. Genetics (in press). (1977).
Cohen, D.: A general model of optimal reproduction in a randomly varying environment. J. Ecol. 56:219-228. (1968).
Darwin, C.: The Autobiography of Charles Darwin. (ed. Nora Barlow). New York: W.W. Norton and Co., Inc., 1958.
Demetrius, L.: Natural selection and age-structured populations. Genetics 79:535-544. (1975).
Epling, C., Lewis, H., Ball, F.M.: The breeding group and seed storage: A study in population dynamics. Evolution 14:238-255. (1960).
Gadgil, M., Solbrig, O.: The concept of r- and K-selection: Evidence from wild flowers and some theoretical considerations. Amer.

Natur. 106:14-31. (1972).

Gains, M.S., Krebs, C.J.: Genetic changes in fluctuating vole populations. Evolution 25:702-723. (1971).

Hamrick, J.L., Allard, R.W.: Correlations between quantitative characters and enzyme genotypes in *Avena barbata*. Evolution 29: 438-442. (1975).

Harper, J.L.: Establishment, aggression, and cohabitation in weedy species. In The Genetics of Colonizing Species. (ed. H.G. Baker and G.L. Stebbins). New York: Academic Press, 1965, pp. 243-268.

Harper, J.L., McNaughton, I.H.: The inheritance of dormancy in inter- and intra-specific hybrids of *Papaver*. Heredity 15:315-320. (1960).

Harper, J.L., White, J.: The dynamics of plant populations. Proc. Adv. Study Inst. Dynamics Numbers Popul., Oosterbeek. pp. 41-63. (1970).

Harper, J.L., White, J.: The demography of plants. Ann. Rev. Ecol. Syst. 5:419-463. (1974).

Jain, S.K.: Patterns of survival and microevolution in plant populations. In Population Genetics and Ecology. (ed. S. Karlin and E. Nevo). New York: Academic Press, 1975, pp. 49-90.

Kahler, A.L., Clegg, M.T., Allard, R.W.: Evolutionary changes in the mating system of an experimental population of barley (*Hordeum vulgare* L.). Proc. Nat. Acad. Sci. U.S. 72:943-946. (1975).

Levins, R.: Dormancy as an adaptive strategy. Symp. Soc. Exp. Biol. 23:1-10. (1969).

MacArthur, R.H.: Some generalized theorems of natural selection. Proc. Nat. Acad. Sci. U.S. 48:1893-1897. (1962).

Mulcahy, D.I.: Gamete Competition in Plants and Animals. Amsterdam-New York: American Elsevier, 1975.

Norton, H.T.J.: Natural selection and Mendelian variation. Proc. London Math. Soc. Series 2:28:1-45. (1926).

Prout, T.: The relationship between fitness components and population prediction in *Drosophila*. I. The estimation of fitness components. Genetics 68:127-149. (1971).

Schaal, B.A.: Population structure and balancing selection in *Liatris cylindracea*. Ph.D. Thesis, Yale University, 1974.

Schaal, B.A.: Population structure and local differentiation in *Liatris cylindracea*. Amer. Natur. 109:511-528. (1975).

Schaal, B.A., Levin, D.A.: The demographic genetics of *Liatris cylindracea* Michx. (Compositae). Amer. Natur. 110:191-206. (1976).

Sharitz, R.R., McCormick, J.F.: Population dynamics of two competing annual plant species. Ecology 54:723-740. (1973).

Tamarin, R.H., Krebs, C.J.: *Microtus* population biology. I. Genetic changes at the transferrin locus in fluctuating populations of two vole species. Evolution 23:183-211. (1969).

Tinkle, D.W., Selander, R.K.: Age-dependent allozymic variation in a natural population of lizards. Biochem. Genet. 8:231-237. (1973).

Ward, R.H., Weiss, K.M.: The Demographic Evolution of Human Populations. New York: Academic Press, 1976.

Williams, G.C.: Sex and Evolution. Princeton, New Jersey: Princeton University Press, 1975.

Williams, J.T., Harper, J.L.: Seed polymorphism and germination. I. The influence of nitrates and low temperature on the germination of *Chenopodium album*. Weed Res. 5:141-150. (1965).

Some Genetic Consequences of Being a Plant

D.A. LEVIN

A. Introduction

During the past decade, great strides have been made in empirical and theoretical plant population ecology. The empirical studies have been highlighted by work on the effect of intraspecific competition on plant size and numbers (Harper and White, 1970, 1974), and on mortality and fecundity schedules (Harper and White, 1970, 1974; Sarukhan, 1974; Sarukhan and Harper, 1973; White and Harper, 1970), the spatial dynamics of succession and colonization (Yarranton and Morrison, 1974), safe site specificity (Harper, 1961, 1965; Sheldon, 1974) pollen and seed dispersal (Levin and Kerster, 1974), and the dynamics and longevity of the seed pool (Harrington, 1972; Roberts, 1972). Within the realm of theory, special attention has been given to optimizing life history strategies with special regard to seed dormancy (Cohen, 1967, 1968), and dispersal (Gadgil, 1971; Roff, 1975), reproductive schedules and developmental switching, (Levins, 1968; Cohen, 1971, 1976; Bradshaw, 1974; Schaffer, 1974a, b, Taylor et al., 1974) and longevity (Gadgil and Bossert, 1970; Gadgil and Solbrig, 1972; Kawano, 1975; Schaffer and Gadgil, 1975).

The descriptions and predictions of life history tactics and responses of species adapted to different environments largely have been considered without regard to the genetic consequences which accrue from the adaptations. Some aspects of demographic attributes which have received theoretical genetic treatment are: population growth with density-dependent regulation (Charlesworth and Giesel, 1972; Charlesworth, 1971, 1973), age-specific selection (Anderson, 1971; King and Anderson, 1971) age-specific fecundity (Giesel, 1974; Demetrius, 1975), and migration (Jain and Bradshaw, 1966; Antonovics, 1968; Gillespie, 1974, 1975, 1976; Bullock, 1976; Nagylaki, 1976). With the exception of immigration, many of the models employed do not take into account the unique properties of plants, nor do they address many interesting questions relevant to plants.

My purpose is to discuss some genetic implications of demographic properties of plant populations. Emphasis will be placed upon the genetic consequences of different fecundity distributions, reproductive schedules, sex ratios, patterns of differentiated plant subdivisions, and seed pool characteristics. Hopefully, this presentation will stimulate others to explore a prime interface between plant population ecology and population genetics.

Effective Population Size

One important parameter which is affected by several demographic features is the stochastic variation in gene frequencies due to random sampling of gametes. To investigate the cumulative effect of sampling error on the genetic constitution of populations, it is convenient to assume a simple (idealized) population structure consisting of N breeding individuals produced each generation by random union of N females and N males extracted as random samples from the gene pool of the previous generation. However, an actual population is apt to have a more complex breeding structure; and it is desirable to reduce such a complex situation to an equivalent simple case for which the mathematical treatment is much easier. The concept of effective population size (N_e) formulated by Wright (1931) meets this requirement. It is the most convenient way of expressing the dispersive process in the form of the variance of gene frequencies or the rate of inbreeding. The effective size of a population is the number of individuals that would give rise to the sampling variance or rate of inbreeding appropriate to the conditions under consideration if they bred in the manner of the idealized population. The effective size typically is less than the actual number of breeding individuals within a population. The concept of effective size can be applied to subpopulations as well as populations.

The fate of a mutant gene is greatly influenced by the effective size of a population. For a neutral gene in a large population, the time to fixation is $4N_e$; extinction time is much shorter than the fixation time and is approximately

$$\bar{t}_o = 2(N_e/N)\log_e(2N) \quad (1)$$

if $\underline{p}$ is $1/(2N)$ (Kimura and Ohta, 1971). Thus, decreasing the effective size reduces both the fixation and extinction times for novel neutral alleles. The fixation time of a favored gene is shorter than that of a neutral one. For an advantageous mutant the probability of fixation is roughly

$$u \approx 2_{s_1}(N_e/N) \quad (2)$$

where s_1 represents the selective advantage, if $4N_e s_1 > 1$ and $2(N_e/N)s_1$ is small (Kimura and Ohta, 1971). We may consider the rate of gene substitution (K) in evolution measured by the average number of gene substitutions per generation. If $\underline{v}_m$ is the number of advantageous mutant genes which appear in the entire population each generation,

$$K = 2s_1(N_e/N)\underline{v}_m \quad (3)$$

Thus, the probability and rate of fixation of a selectively advantageous mutant is an inverse function of effective size.

B. Fecundity Distribution in Plants

Recent studies on the population ecology of plants have shown that in natural and experimental populations only a small proportion of the seedling crop survives to reproductive maturity, and that a size hierarchy will be established among the seedlings early in the season which will persist and be magnified as the population matures (White and Harper, 1970; Ross and Harper, 1972). As a consequence, populations typically are characterized by biomass and fecundity distributions which are L-shaped, wherein a very small proportion of the population may make a substantial and grossly disproportionate contribution to the yield or seed output of the population (Koyama and Kira, 1956; Risser, 1969; Sarukhan, 1974; Leverich, 1977). Indeed, in some populations, the highest fecundity classes may be more frequent than plants of medium fecundity. Conversely, a very large fraction of the population may make a small contribution to yield or seed output. The difference in seed production between large and small plants within a population may differ by several orders of magnitude, especially in species with great developmental flexibility. Even the difference between mean seed set and seed set in individuals with the most reproductive output may be enormous (Salisbury, 1976). The L-shaped performance distribution may arise from competitive interactions between cohorts and from the statistical distribution of site quality, as determined in part by the genetic makeup of the population.

The typical fecundity distribution in plant populations is in stark contrast to the Poisson distribution which is a cornerstone of population genetic theory (Karlin and McGregor, 1968). What are the genetic consequences of an excess of small plants and an excess of large ones relative to a Poisson distribution?

The effective size of a population may be defined by its fecundity or progeny distribution. Wright (1938) showed that if N parents furnish varying numbers (k) of gametes to the next generation, and if the population is stationary in size (k=2), and with panmixia, the effective size is

$$N_e = \frac{4N - 2}{V_k + 2} \qquad (4)$$

where V_k is the variance of k. If the progeny numbers follow the Poisson distribution (V_k = k=2), the effective size becomes roughly equal to the actual number; i.e., N_e = N. If, as in most populations, the variance is larger than the mean, the effective size becomes less. When differences in fecundity are hereditary, as opposed to representing different plastic responses to varying site hospitality, the effective size becomes

$$N_e = \frac{4N}{(1 + 3h^2)V_k + 2} \qquad (5)$$

where h^2 is the heritability of progeny numbers (Nei and Murata, 1966). Therefore, the effective size decreases as the heritability of fitness differences increases.

The effect of different progeny distributions are best conveyed with a series of examples. Consider a population composed of 100 plants which is stationary in size so that the average contribution of gametes to the next generation is 2. If the gamete contribution were following a Poisson distribution, the effective size would be 100. As the gamete distribution becomes more L-shaped, the effective size declines as shown in Figure 1. When most plants have no offspring, the effective size will be less than half that of the number of adult plants. In the last example, 60% of the plants contributed no gametes, and a small fraction make a disproportionately large contribution to the next generation. The N_e is 30.6. In view of the very low probability of a seed escaping predation or disease, germinating, and giving rise to an adult plant (Harper and White, 1974), the number of plants in nature falling in the zero class probably is much larger than in the last example, as would be the variance in reproduction of the adult plants. Thus, the deviation from a Poisson progeny distribution is

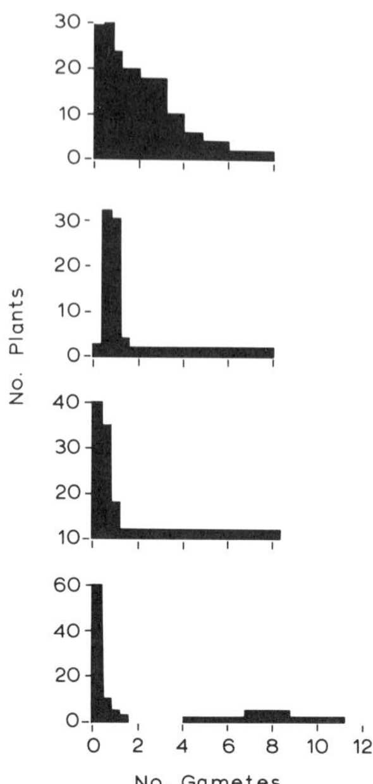

Figure 1. The effective size of populations of 100 plants based upon 4 fecundity distributions.

probably much greater than shown here; and, accordingly, the N_e is much smaller. The difference in N_e in the examples was the result of chance differences in fecundity. If selection in part dictated where plants (genotypes) were positioned on the fecundity distribution, the effective sizes would be smaller than shown in Figure 1.

The fecundity distribution of plant populations is dependent on the nature of the response to density stress from conspecifics or to other biotic and abiotic environmental pressures. Species exhibit plastic (developmental) and mortality responses, the balance between the two depending upon the species and the circumstances. When the balance is in favor of the former, fecundity hierarchies like those described in Figure 1 are likely to develop. When the balance is in favor of the latter, most plants will not reproduce, and a few will be relatively fruitful. Also, there may be a gross discontinuity between the zero class and those classes in which the reproductive individuals reside. A mortality response will tend to yield a higher variance in fecundity than a plastic response, since the zero class and the mean fecundity probably will be much larger than in the former.

As noted earlier, the decay of genetic variability and the response of populations to selection are a function of their fecundity distributions. As the distributions become more L-shaped, and correlatively as the effective size decreases, the time to extinction or fixation of a novel neutral mutation declines as does the ability of populations to retain polymorphism. Thus, we may expect populations balanced in favor of a plastic response to be more variable than those balanced in favor of a mortality response. Heritability of fecundity differences further accelerates the dispersive process for most of the genome, since the effective size is reduced. Smaller effective size is accompanied by more rapid response to selection for the characters responsible for the fecundity differential. Thus, populations with emphasis upon mortality response may experience more rapid selective gene frequency change than those with a plastic response. Indeed, selection in the former populations is similar to truncation selection in which only a small superior fraction of the population constitutes the breeding component (Crow and Kimura, 1970). The effect of plasticity and mortality on the ability of populations to respond to selection and maintain polymorphism is summarized in Figure 2.

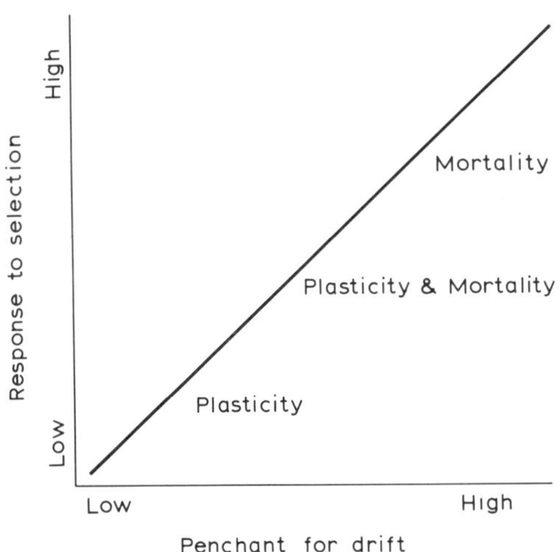

Figure 2. The effect of plastic and mortality responses on the evolutionary potential of populations.

C. Disproportionate Sex Ratios

Sex ratios in dioecious plants not infrequently depart significantly from unity (Lewis, 1942; Godley, 1964; Mulcahy, 1967; Dzhaparidze, 1963; Lloyd, 1974; Opler and Bawa, 1977). The departure may be in favor of males or females as seen in Table 1. The ratios presented there are averages, and may vary from population to population as a function of the habitat (Gatina, 1955; Il'in, 1969; Freeman et al., 1976) or degree of crowding (Mulcahy, 1967; Harris, 1968). In some species, certain populations have substantial male excesses whereas others have female excesses. In the British Isles, males of *Potentilla fruticosa* were most abundant away from some streams, whereas females prevailed along streams (Richards, 1975). In the area of the Lower Volga in Russia, the sex ratios of *Fraxinus lanceolata* and *Acer negundo* are dominated by males; but in more mesic environments females are as common as males (Lysova and Khizhnak, 1975). In the western United States, the sexes of *Thalictrum fendleri*, *Ephedra viridis*, *Acer negundo*, *Distichlis spicata* and *Atriplex confertifolia* segregate along moisture gradients (Freeman et al., 1976).

Disproportionate ratios occur in spite of the advantage which theoretically accrues to a 1:1 ratio (Fisher, 1930; Kolman, 1960; Bodmer and Edwards, 1960; Leigh, 1970; Eshel, 1975) or a slight female excess in plants (Kaplan, 1972; Lloyd, 1974). The adaptive significance of distorted ratios remains to be determined. The mechanisms by which the distortions are, or may be, achieved are better understood. These are: gametophyte selection (Godley, 1964; Lloyd, 1974), differential maturation and mortality (Clark and Orton, 1967; Lloyd; 1973; Spieth, 1974), differential proclivity of sexes to reproduce vegetatively (Putwain and Harper, 1972), and developmental responses to stress (Heslop-Harrison, 1972).

Wright (1931) has shown that if a population consists of N_m males and N_f females, the effective size is

$$N_e = \frac{4N_mN_f}{N_m + N_f} \qquad (6)$$

Unless the number of the two sexes is equal, the effective size is always smaller than the actual numbers of males plus females. Especially, when the number of males and females differ greatly N_e depends mainly on the minority sex.

Consider the effective size of populations of 100 plants in some of the species noted in Table 1. In *Anistome aromatica* with a sex ratio of 1:5 the effective size would be about 42. In *Humulus lupulosis* and *Rumex thyrsiflorus* with sex ratios of 1:9, the effective sizes are about 36. As the differential between the sex becomes less pronounced, the impact on the effective size is greatly diminished. For example, *Clematis paniculata, Coprosma rhamnoides,* and *Randia subcordata* have sex ratios of about 2:1. The effective size in their populations would be about 88. Because of their impact on the effective size of populations, disproportionate sex ratios decrease the time to fixation for a novel neutral mutation, and increase the probability and rate of fixation of selectively advantageous mutants.

D. Reproductive Format and Population Stability

The reproductive schedules of plants are extremely diverse. Some plants flower the first year of their life and then die; others flower periodically after a pre-reproductive period of two to several years; and others only flower once after achieving considerable age (Harper and White, 1974). Correlatively, the generation times of different species may vary by at least an order of magnitude.

The effect of divergent reproductive schedules on the response of populations to selection has been addressed by agronomists and theoretical geneticists. It is well known that the rate of response to directional selection in chronological time is inversely related to generation time (Falconer, 1960). For example, if the generation time is 4 years in one population and 1 year in another, the rate of gene frequency change per year will be 4 times greater in the second population. It follows that when the direction of selection changes in a random or cyclic fashion, populations with long generation times are less able to track these changes than populations with short generation times. The result is that the genetic structure of populations with long generation times are likely to be tuned to

Table 1. Examples of disproportionate sex-ratios in angiosperms.

Species	Ratio ♂/♀	Reference
Silene otites	1.63	Lewis, 1942
Cannabis sativa	0.72	"
Humulus japonicus	0.40	"
Humulus lupulosus	0.11	"
Rumex acetosa	0.40	"
Rumex thyrsiflorus	0.11	"
Clematis afaliata	2.13	Godley, 1964
C. paniculata	2.00	"
Aristotelia serrata	2.33	"
Melicytus ramiflorus	2.60	"
Cyathodes colensio	3.00	"
Coprosma rhamnoides	2.00	"
Macropiper exelsum	4.25	"
Aciphylla monroi	2.42	Lloyd, 1973
Anistome aromatica	7.50	"
Cordia collococea	1.50	Opler & Bawa, 1977
Cordia panamensis	2.11	"
Erythroxylon rotundifolium	2.14	"
Coccoloba caracasana	0.41	"
Triplaris americana	0.71	"
Allophyllis occidentalis	1.48	"
Albertia edulis	1.69	"
Randia spinosa	2.70	"
Randia subcordata	2.04	"
Zanthophylum setulosum	1.92	"
Thalictrum revolutum	2.45	Levin, unpublished
Thalictrum fendleri	0.24 & 6.77[a]	Freeman et al., 1976
Distichlis spicata	0.59 & 2.96[a]	"

[a]Minor sex varies with habitat.

long-term environmental pressures, and retain a relatively high level of genetic polymorphism. The genetic structure of populations with short generation times may be more closely tuned to recent environmental pressures, but the cost would be a reduction in polymorphism.

The effective size of populations is also a function of the life history attributes of populations. The equation for effective size when the population size is stationary is

$$N_e = N_m \underline{t} \tag{7}$$

where N_m is the number born per year who survive until the mean age of reproduction, and $\underline{t}$ is the mean age of reproduction or generation time (Nei and Imaizumi, 1966). Thus, the shorter the generation time, the smaller will be the effective size, given that the number of plants per cohort surviving until the mean reproductive age is unaltered. This suggests that a shift from the perennial to annual habit (all other things being equal) might be accompanied by a more rapid decay in genetic variability per population and greater gene frequency variance between populations. Returning to perennials, the aforementioned formula indicates that the survivorship curve is important in dictating effective size, since the mean age of reproduction may be established by very different survivorship distributions. If most reproductive plants have relatively short lives whereas a few are very long-lived, the effective size would be lower than if the longevities were less disparate.

The effective size of populations is also a function of systematic or random changes in population size. If a population is expanding in size, the effective size also will expand, although the rate of increase in the latter will be much slower than that of the standing crop. Likewise, contracting populations decline in effective size. Of particular importance is that a population fluctuating about some given size has a smaller effective size than would a population whose number were constant through time. Wright (1938) showed that if periodic reductions occur in the number of breeding individuals, the effective size is the harmonic mean (reciprocal of the mean of the reciprocals) over some time span. Since the harmonic mean is dominated by small numbers, cycles in population size greatly depress the effective size. For example, if a population of annual plants expands in five generations from 10 to 1,000,000 in geometric series, and then returns to the initial size through the following five generations, the effective size would be about 54. Note that reduction in the number of breeding individuals does not have to be accompanied by a decline in population size. In a population of perennials, stable in age structure and stationary in numbers, the variation in the percentage of adult plants flowering will also depress effective size. Such variation is the rule rather than the exception (Harper and White, 1974).

Flowering plants are either self-incompatible or self-compatible. The evolution of self-compatibility from self-incompatible progenitors has occurred many times in many plant groups (Stebbins, 1957, 1958; Grant, 1958, 1971; Jain, 1976). Self-compatible plants may have a prepotency for pollen from other plants and thus be predominantly cross-fertilizing, or plants may be predominantly self-fertilizing. The level of outcrossing, rather than the breeding system per se, is of special importance. This is because it determines the patterns in which gametes are united and is thus a prime determinant of the genotypic frequency distribution. The level of outcrossing also determines the tempo at which existing gene assemblages are broken up and new ones are produced (Allard, 1975). For a two-dimensional population, a population with areal continuity, the effective size of panmictic units or neighborhoods within populations may be estimated as:

$$N_e = \frac{12.6\sigma^2 d}{1 + F}; \quad (8)$$

where σ^2 is the variance of pollen and seed dispersal curves, d is the density of plants, and F is the fixation index related to the outcrossing levels (Wright, 1946). As selfing increases, neighborhood size decreases, and the potential for gene frequency heterogeneity between subpopulations increases, as does isolation by distance.

The level of outcrossing for several self-compatible species is listed in Table 2. It is evident that self-compatibility does not provide a basis for drawing inferences about the mating system; outcrossing rates vary over a broad range. Not only does outcrossing vary between species, it varies widely within a species from one population to another in time and space (Allard et al., 1968 and references in Table 2). For example, in natural populations of *Lupinus succulentus*, the level of outcrossing varies from 13% to 97%. Wide fluctuations in this parameter from year to year permit the population to maintain

Table 2. Estimates of Outcrossing in Self-Compatible Species.

Species	Mean Percentage Outcrossing	Reference
Festuca microstachys	< 1	Allard & Kahler, 1971
Trifolium subterranean	< 1	Marshall & Broue, 1973
Hordeum jubatum	1	Babbel & Wain, 1977
Avena barbata	2	Jain, 1975
Plectritis brachystemon	2	Ganders et al., 1977
Avena fatua	3	Jain, 1975
Avena sterilis	4	Jain, 1975
Hordeum vulgare	4	Jain, 1975
Bromus mollis	4	Jain, 1975
Trifolium hirtum	5	Jain, 1975
Medicago polymorpha	8	Jain, 1975
Collinsia sparsiflora	10	Allard & Kahler, 1971
Lycopersicon pimpinellifolium	14	Rick et al., 1977
Phlox cuspidata	22	Kahler & Levin, unpubl.
Senecio vulgaris	22	Campbell & Abbott, 1976
Papaver dubium	25	Humphreys & Gale, 1974
Lupinus succulentus	46	Harding & Barnes, 1977
Lupinus nanus	54	Harding et al., 1974
Clarkia exilis	55	Vasek & Harding, 1976
Clarkia tembloriensis	57	Vasek & Harding, 1976

a higher level of heterozygosity than would be suggested by the mean value (Jain, 1975). Intraspecific variation in outcrossing may be due to differences in plant density and the attendant behavior of pollinaors or receipt of wind-borne pollen. Flower structure, habitat hospitality as it affects plant density and flower structure, and the genetic composition of the population also influence intraspecific outcrossing rates.

In some species, mixed selfing and random mating may be balanced by the crossing behavior of different genotypes. For example, in *Borago officinalis*, homozygotes are obligate outbreeders and heterozygotes are self-compatible (Crowe, 1971). Similarly in *Vicia faba*, the products of selfing tend to outcross and vice versa (Drayner, 1956, 1959; Holden and Bond, 1960). In other species the breeding system may be in transition. Kahler et al., (1975) have described the evolutionary changes in the breeding system of an experimental population of *Hordeum vulgare*. The outcrossing rate more than doubled during the twenty generations spanned by the study; the change was due to heterozygote advantage. The breeding system may be altered in a systematic fashion through selection of the floral architecture which influences the proximity of the stigma and style, as has been demonstrated in *Nicotiana rustica* (Breese, 1959).

The breeding system in many flowering plants is responsive to environmental cues, which are likely to change the level of outcrossing. The plastic response of the breeding system is best exemplified in species with both chasmogamic (adapted for cross-pollinating) and cleistogamic (self-pollinating prior to opening) flowers. In the grass *Stipa leucotricha*, chasmogamy is the norm, but cleistogamy is triggered by unfavorable conditions of soil or moisture supply (Brown, 1952). In the facultatively cleistogamic grass *Bromus carinatus*, chasmogamy is stimulated by favorable growing conditions (Harlan, 1945). In another grass, *Rottboellia exaltata* (Heslop-

Harrison, 1959), and in *Lespedeza cuneata* (McKee and Hyland, 1941) the incidence of cleistogamy is under photoperiod control. In *Lithospermum carolinense*, the percentage of plants with at least some cleistogamic flowers increases with plant density (Levin, 1972). In this species some populations produce only outcross seed, while others produce less than 60% outcross seed. Some species produce cleistogamic flowers in the beginning and the end of their flowering seasons, and chasmogamic ones during the remainder of the flowering season (e.g. *Viola odorata*, Madge, 1929; *Lamium amplexicaule*, Bernstrom, 1950). Other species such as *Senecio vulgaris* produce cleistogamic flowers only at the beginning of the season (Haskell, 1954). Environmental influence on the breeding system need not be manifested in gross differences in floral morphology. In *Myosurus sessilis*, the self-pollination occurs during the first part of the year and outcrossing during the later part (Stone, 1959). The potential for cross-pollination, which is dictated by the proximity of the stamens and pistil is greatly increased when there is plenty of water. In *Hordeum vulgare*, which is predominantly self-pollinating, the advent of frost increases the level of cross-pollination (Suneson, 1953). In summary, the developmental responses of plants to environmental pressures and the responses of plant production to the same or other exigencies result in considerable heterogeneity in the breeding structure of populations in space and time.

Thus far, our discussion has assumed that reproduction is sexual. However, hundreds of species are facultative apomicts capable of reproducing in part by agamospermy (unfertilized seed) or by vegetative structures (stolons, rhizomes and bulbils). Agamospermy is more common in fugitive species, and vegetative reproduction is more common in late successional or climax species (Gustaffson, 1946; Nygren, 1954; van der Pijl, 1969; Salisbury, 1976). Facultative apomixis affects two parameters discussed earlier, generation time and the fecundity schedule. Both agamospermy and vegetative reproduction increase the generation time, since the lives of individuals (genets) are extended, although the distribution of a genet in space will depend upon the mode of asexual reproduction. Accordingly, partially asexual reproduction will increase the effective size of a population and reduce the rate of response to selection in chronological time. Thus, they might be better buffered against the decay of variability than sexual populations. However, the fecundity differential among genets which can reproduce asexually may be much more dramatic than among genets capable only of sexual reproduction, because fecundity will be a compound function of genotype, habitat hospitality, the growth rate of the genet, and the mechanism of asexual reproduction. Since increasing the variance in progeny numbers reduces the effective size of a population, especially if differential fecundity is heritable, the consequences of extended generation time may be counterbalanced and even overwhelmed by the L-shaped fecundity distribution.

For the purpose of demography, one may treat all plants as individuals regardless of their genetic affinities. However, for genetic purposes, the number of genets (genotypes) is an important parameter, the number of plants notwithstanding. Given a moderate, fluctuating population size, we might expect populations of facultative apomicts to be genetically depauperate. This does not conflict with the idea of a slow rate of decay of genetic variability based upon a relatively large N_e. Moreover, we may expect these plants to display considerable heterogeneity in the relative frequencies of specific genets from one population to another. If the plants are self-incompatible, as is typical of facultative apomicts, and reproduce vegetatively, then the mean distance from which pollen comes for the production of sexual seed increases, as the size of clones increases. The result is to increase the neighborhood area, and thus reduce the potential for local population subdivision fostered by the asexual component of the reproductive system (Levin and Kerster, 1971).

Vegetative and sexual reproduction place conflicting demands for energy upon a plant, and one mode of reproduction is achieved at the expense of the other (Gustafsson, 1946-1947; Harper, 1977). Accordingly, when vegetative reproduction is extensive, flowering is likely to be modest or absent. Vegetative reproduction is density-dependent; increasing plant density shifts the balance between asexual and sexual reproduction toward the former (e.g., *Frageria virginiana*, Holler and Abrahamson, 1977; *Rubus hispidus*, Abrahamson, 1975; *Tussilago farfara*, Ogden, 1974; *Hieracium floribunda*, Thomas, 1974). Interference from other species also make a pronounced impact on this balance as demonstrated in *Agropyron repens* (Sagar and Mortimer, 1977), and *Solidago canadensis* (Werner, personal communication). In the latter the percentage of biomass allocated to sexual reproduction increases and that allocated to vegetative reproduction decreases as one goes from a garden to an early successional community to a late successional community. The formation of bulbils also is influenced by the environment. In *Poa bulbosus* the formation of bulbils and florets vary as a function of light and temperature, short days and low temperatures triggering the development of bulbils (Younger, 1960). Short days stimulate the development of bulbils in *Cynosurus cristatus* (Wycherly, 1954), and *Deschampsia caespitosa* (Nygren, 1949); long days have the same effect in *Kalanchoe daigremontiana* (Roberts and Struckmeyer, 1938).

The level of agamospermy in facultative apomicts also varies with the environment. Saran and deWet (1970) studied a clone of *Bothriochloa intermedia* that reproduces sexually in the summer, but is an obligate apomict in the winter. With a 14 hour day length this biotype was highly sexual, but with a 12 hour day length 43% of the embryo sacs were unreduced, i.e., they would produce agamospermic seed. The incidence of agamospermy also varies with the season in *Arabis holboellii* (Bocher, 1951), and *Calamagrostis purpurea* (Nygren, 1951). In *Dichanthium aristatum* the proportion of unreduced embryo sacs is affected in a systematic manner by the light regime (Knox and Heslop-Harrison, 1963). Under continuous exposure to short days, up to 79% of the embryo sacs are unreduced compared to 47% when plants are exposed to the minimum photoinductive regime of 40 short days. Temperature has a pronounced effect on the formation of unreduced embryo sacs in *Limonium transwallianum*, *Oxybaphus nyctagineous* and *Marabilis jalapa* (Hjelmqvist and Grazi, 1964, 1965).

The adaptive calculus of facultative apomicts may involve a developmental response to environmental cues which may or may not be associated with stress. The plastic responses of the reproductive system of these plants may be much greater than the mean differences between species, and thus must be taken into account when assessing the genetic consequences of facultative apomixis. It seems that during periods of stress, sexual reproduction is favored. The extent to which plastic responses and the evolutionary tuning of the genetic system of species are in accord or conflict is poorly understood. Plastic responses may provide a hedge against the evolutionary bet.

E. Seed Pool Dynamics

Thus far, rates of evolution have been considered with regard to life history features of perennial plants. Annuals should be considered as well since they too have age-structured populations by virtue of their seed pools. The annual habit in plants is accompanied by specialized physiological mechanisms which permit seeds to remain dormant in the soil for a few years to decades (Koller, 1969; Mayer and Poljakoff-Mayber, 1975; Roberts, 1972). The nearly universal imposition of germination regulating mechanisms prevents the reproductive potential of a population from being gambled except when the probability of survival is favorable. The longevity of seeds tends to be a positive function of environmental unpredictability and is correlated with other adaptations involved with coping with unpredictability. The seeds of desert ephemerals and fugitive species are relatively long-lived (50 to 100 years), whereas the seeds of mid-successional annuals in mesic habitats often are short-lived (10 to 20 years) (Toole and Brown, 1946; Harrington, 1972; Kivallaan and Bandurski, 1973). The buried seed population has a constant death risk (Harper and White, 1974).

The seed bank is based upon contributions made over several years, during which time selection may have favored alternate alleles at particular loci related to temperature tolerance, pubescence, etc. The seed pool therefore provides the population with a memory of its recent history and is tuned to the general experience of the population over several years rather than just the past one or two. A seedling crop is drawn from this potentially diverse seed pool.

The optimal germination tactics for a seed pool have been studied by Cohen (1966, 1967, 1968). The question now is not what is best for a population in terms of its persistence, but what are some genetic consequences of different seed pool histories? What constraints do seed pools impose on evolution at a single locus unrelated to dormancy? Alan Templeton and I (unpublished) have considered the effects of directional and disruptive selection in time, taking into account the impact of the environment on population size. I will summarize some of our findings which are based upon analytical solutions.

By virtue of its history, the seed pool retards the rate of evolution when fitness is constant, balancing or directional selection notwithstanding. There is a simple linear relationship between the change in gene frequency per generation and the average number of years a seed has spent in the seed pool prior to germination. Consider a case of directional selection in a population of annuals. If the mean time that plants spent in the seed pool prior to germination were 10 years, then the number of years required for selection to alter the frequency of a gene from one value to another would be 10 times greater than if there were no seed pool. If the average time in the seed pool were 3 years, then the rate of evolution would be 3 times greater than if there were no seed pool. Consider a case of balancing selection at a single locus where the heterozygote was favored and both homozygotes had relative fitness of .99. The number of years required to move a gene from one frequency to the equilibrium value (.5) also would be a simple function of time plants spent in the seed pool.

The extent to which the seed pool retards the response to selection not only is dependent upon the age structure of the seed pool, but also is dependent upon the growth rate of the population. The discussion thus far has assumed that population size is constant. Should populations pass through a period of growth, the seed pool would be weighted in favor of recent seed since it would be increasingly abundant and constitute a higher proportion of the pool than if the contribution was not accelerating. The mean age of the seeds in the pool would decline. On the other hand, if populations were contracting, the seed yield per year would also contract so that the pool would be weighted in favor of the older seeds relative to what would be the case with a stable population size. The mean age of seeds in the pool would increase. Accordingly, the rate of evolution will be faster in expanding populations with seed pools than in stationary populations. Correlatively, the rate of evolution will be slower in contracting populations than in stationary ones.

The seed pool allows the existence of a covariance between fitness and absolute seed production during years in which the environment is hospitable and plants generally are large. The genotype favored in that environment will produce more seeds and thus have a greater impact on the seed pool than an alternate genotype favored in years when conditions for survival are marginal. With a seed pool not only is relative fitness important, but the absolute performance of the best genotype in year one versus the best in year two is important as well. Differential contribution of seeds to the seed pool both within and between years, especially the latter, may have the same consequences as directional selection. Let us assume that in good years the relative fitnesses of three genotypes are: AA = 1.00, Aa = 0.75, and aa = 0.50. Let us also assume that in marginal years the fitnesses are: AA = 0.50, Aa = 0.75, and aa = 1.00. Population size is constant. In the good years, the seed production of the favored homozygote (AA) is 100, whereas in the marginal years the seed production of the favored homozygote (aa) is only 50. If the environment alternates between good and marginal years in a regular cyclic fashion, the A gene eventually will be fixed even though environments and relative fitnesses (within years) are symmetrical. Random fluctuations in the environment will have the same effect. Accordingly, selection will favor genes that do the best when the environment is hospitable. The greater the mean age of seeds in the pool, the more pronounced is the filtering effect of selection in a fluctuating environment.

The memory of past environments is not restricted to annuals. Perennial plants also have a memory, but the information is stored above the ground in the growing portion of the life cycle instead of the dormant one. Consider the case of the perennial which is long-lived and reproduces each year until it dies. Assume that selection operates on a juvenile characteristic and that after the first year the genes responsible for juvenile expressions are no longer expressed. After the first year, the genes for juvenile characters are stored in a non-selective manner by the adult plants. If the environment changes, and the phenotype which conferred maximum fitness upon seedlings when the adult plants were young is no longer optimal, the adult plants will still continue to transmit the outmoded expression. Indeed, how can it be otherwise? As a consequence, the evolutionary constraints imposed by a seed pool also are experienced by perennials exposed to age-specific selection. The extent to which perennial populations respond to selection is dependent upon the longevity of reproductive individuals, as well as their age-specific fecundity. Many perennials, especially those which are woody, or herbs which spread by stolons and rhizomes or disperse by bulbils or apomictic seed, increase their seed output with age. Accordingly, if the environment is changing in a systematic, directional fashion, the plants contributing the most seed are those least capable of leaving fit offspring. The memory in perennials buffers the populations against radical genetic change in temporarily fluctuating environments. Under some circumstances the memory also permits directional evolution in the face of these environments, as is the case with seed pools.

F. The Effect of Patch Size and Shape on Migration-Selection Equilibria

There has been much interest in the joint action of migration and selection on the level of genetic variability in subdivided populations occupying spatially heterogeneous environments. Two- and multi-niche models have been constructed which relate to organisms in general (Christensen, 1975; Karlin and Kenett, 1977; Gillespie, 1975, 1975, 1976; Hedrick et al., 1976; Karlin, 1976, 1977) or plants specifically (Jain and Bradshaw, 1966; Antonovics, 1968; Nagylaki, 1976; Levin and Kerster, 1975; Gleaves, 1973). Unfortunately, most treatments make assumptions about the mating system, the spatial distribution of patch types, and dispersal distributions which are contrary to what one actually finds in plant populations. Plant populations are not panmictic; incoming genes are not randomly distributed throughout a population; and environmental patches are two-dimensional arrays with specific dimensions.

Using computer simulation, John Wilson and I chose to answer the following questions: What are the equilibrium gene frequencies in subpopulations adapted to different environments when patch size and shape are variables? To what extent do the equilibria differ with different gene dispersal schedules? Some of our results are summarized below. For additional explanation of the model, the reader is referred to Wilson and Levin (1979).

The Model

Consider a population of 4096 plants occupying a site with patch types X and Y in equal proportions, and a single locus with two alleles (A and a). The relative fitnesses of genotypes in patch X are: AA = 1.0., Aa = 0.75, aa = 0.50. The relative fitnesses of genotypes on patch Y are: AA = 0.50, Aa - 0.75, aa = 1.0. The size and shape of patches varies with at least 4 plants per patch. The plants are assumed to occupy a 64 X 64 grid of uniformly-spaced safe sites; every plant is one map unit from its nearest neighbors.

The initial population is composed of AA plants in the X patches and aa plants in the Y patches, with patches alternating in space in a checkerboard fashion. The breeding structure of the population is defined by the dispersal of pollen, and dispersal follows random, leptokurtic or nearest-neighbor patterns. No seed dispersal is assumed, and seeds occupy the same site as the seed parent. The stepping-stone schedule involves the movement of pollen to one of the four or eight nearest sites. Each plant produces 18 pollen grains, and pollen broadcast is random in direction. Plants are assumed to be self-incompatible.

Competitive selection occurs when more than one seed is produced and deposited per site. The surviving seed genotype is a function of the frequency of the genotypes and the relative fitness of each.

Results

The mean frequency of the maladapted gene in each patch type is a function of patch size (Fig. 3). With a leptokurtic pollen distri-

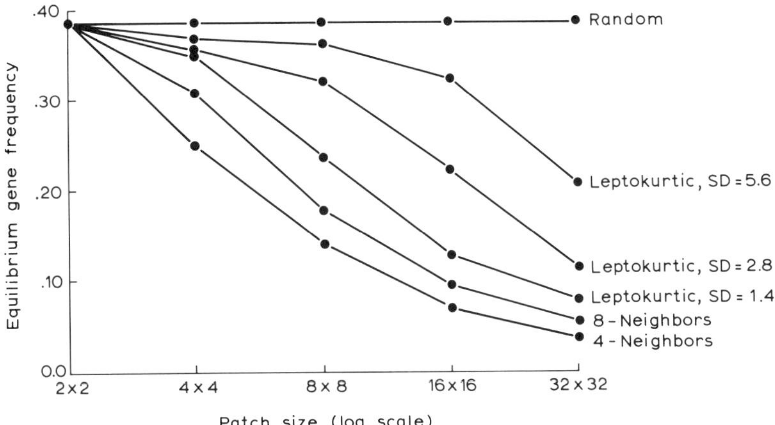

Figure 3. The effect of patch size on the mean maladapted gene frequency.

bution (S.D. = 2.5 units), equilibrium frequencies range from .38 for 2 X 2 patches arranged in a checkerboard fashion to 0.11 for 32 X 32 patches. Moving to a random patch distribution results in lower equilibrium values. Equilibrium gene frequencies are greatly influenced by patch shape. With a patch size of 64 plants, equilibrium frequencies of the maladapted gene vary from .16 in 8 X 8 patches to 0.25 in 4 X 16 patches to 0.34 in 2 X 32 patches and 0.38 in 1 X 64 patches (Fig. 4). These values are based upon checkerboard patch patterns.

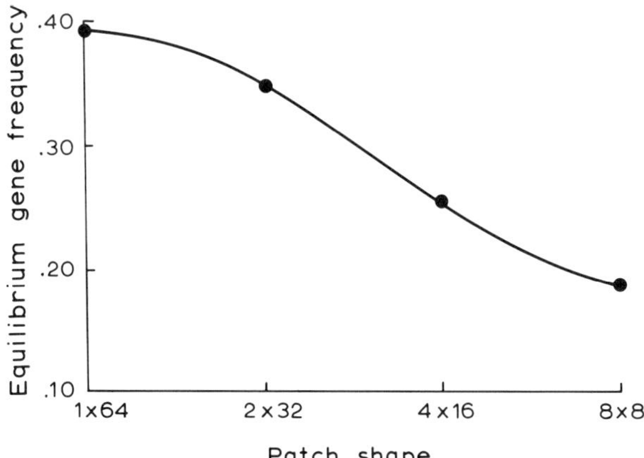

Figure 4. The effect of patch shape on the mean maladapted gene frequency.

The maladapted gene is distributed in an non-random fashion within patches. This is most evident in large patches (Figure 5). This

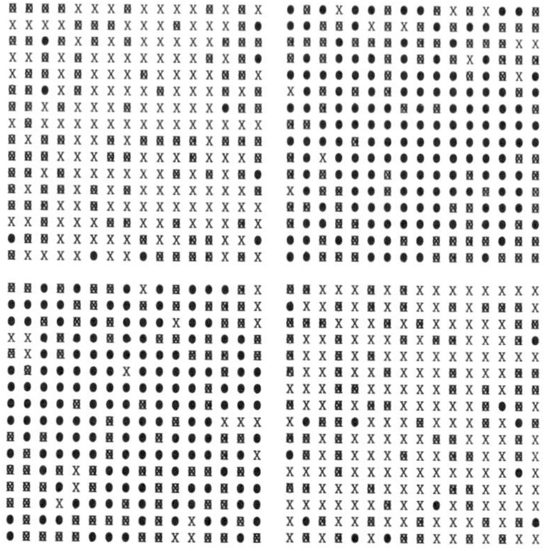

Figure 5. A representative distribution of genotypes within 4 patches after a selection-migration equilibrium. The upper left and lower right patches are of the X type, and the upper right and lower left of the Y type.

gene tends to be higher in frequency near the edge of the patch than near the interior, and in patches larger than 8 X 8 it rarely reaches the center. This pattern results from narrow pollen dispersal between and within patches. The former dictates the initial deposition sites; and the latter, the penetration rate. Once in a patch, an alien gene will move over short distances which are random with regard to direction. This provides ample opportunity for its selective elimination beyond the area of continual immigration.

Thus far, equilibrium frequencies have been considered within the context of a leptokurtic dispersal schedule. It is also of interest to determine the consequences of various schedules including those assumed in most population genetic models. Thus, we compared the equilibrium levels of maladapted genes under the following pollen dispersal schedules: (a) random, (b) leptokurtic with a mean distance of 2.5 and S.D. of 2.8 units as used previously, (c) leptokurtic with a mean distance of 1.25 and S.D. of 1.4 units, (d) leptokurtic with a mean of 5.2 and S.D. of 5.6 units, (e) 4 nearest-neighbors, (f) 8 nearest-neighbors. The leptokurtic distributions had the same kurtosis. The nearest-neighbor distributions provided for equal dispersal of pollen for each of the nearest-neighbors in question.

The effect of the six dispersal schedules on equilibrium frequencies in populations with different patch sizes is shown in Figure 3. The equilibria vary widely, especially in populations with large patch sizes. With random mating, the maladapted gene frequency averages

0.38, patch size notwithstanding. If we consider the 32 X 32 patch population, the frequency of the maladapted gene equilibrates at about 0.40 with pollen dispersal to the four nearest-neighbors, and at 0.07 with pollen dispersal to the eight nearest-neighbors. With narrow, moderate and broad leptokurtic dispersal, frequencies are 0.08, 0.11, and 0.21, respectively. The more restricted the pollen flow, the lower the equilibrium value. Correlatively, the more restricted the pollen flow, the more highly adapted are subpopulations to their respective environments.

In view of the effect of different dispersal schedules on equilibrium gene frequencies, it is of interest to determine the time to equilibrium. With panmixia we would expect subpopulations to reach equilibrium in one generation. With a leptokurtic distribution (S.D. = 2.5 units), equilibria are reached in an average of 12 generations when patches are 8 X 8, or larger. For populations with 2 X 2 patches, equilibria are achieved in an average of 5 generations and with 4 X 4 patches in an average of 7 generations.

Thus far, we have considered patches to be contiguous. However, it is possible that distinctive patches may be separated by small uninhabitable areas. We explored the effect of habitat discontinuity by introducing a 2-position (unit) gap between patches. As one might anticipate, habitat discontinuity and restricted gene flow (in the form of the previous leptokurtic schedule) confer an excellent opportunity for interpatch differentiation. In populations where patch size is large, the frequency of the maladapted gene may be nearly 50% of that in the absence of the gap. The impact of the gap is even larger with more restricted dispersal schedules, since the size of the gap is proportionally greater, relative to the dispersal limits. With nearest-neighbor dispersal, the gap cannot be crossed, and subpopulations remain monomorphic.

In summary, the level of polymorphism within subpopulations occupying heterogeneous environments is a complex function of patch size, shape, spatial distribution of patch types, and gene dispersal schedule. Simply to note that an environment is heterogeneous and that gene flow is restricted fails to provide a meaningful insight into gene frequency heterogeneity between subpopulations, the correlation between patch type and gene frequencies, or the extent to which the adaptedness of populations is reduced by a selection-migration equilibrium. We need to know about the organization and size of patch types if we wish to appreciate the potential and actual level of differentiation which may ensue in the face of gene flow.

G. Conclusions

In 1972, Bradshaw wrote an essay on some evolutionary consequences of being a plant. He emphasized the implications of the strength and diversity of selection which operate on closely adjacent populations, the restriction of gene flow between neighboring populations, the correlation of habitats between parents and offspring, the flexibility of the breeding system, and the nature and organization of genetic variation within species. He concluded that "to understand what is actually happening in plant species, we need to assume very different premises from many of those exercising the minds of many population geneticists." As I have attempted to convey in this presentation, we vigorously endorse this position. Moreover, it is paramount that we explore the evolutionary implications of demographic features which typically have resided in the domain of ecologists. Fecundity schedules, reproductive schedules and mode, seed pool properties, and the pattern of differentiated population subdivisions all are variables which have manifest genetic consequences. These in turn may free or restrict the evolutionary potential of populations. Moreover, the aforementioned demographic properties either have not been considered by population geneticists or their actual expression is contrary to many assumptions used in genetic models. The theory and illustrations presented here only represent a small sample of the kinds of problems that we need to address if we hope to understand what is actually happening in plant species, why it is happening, and what it means.

Acknowledgements. This study was supported by the National Science Foundation Grant No. DEB76-19914. The author is indebted to Henry E. Schaffer for criticizing an early version of the paper.

References

Abrahamson, W.G.: Reproductive strategies in dewberries. Ecology 56: 721-726. (1975).

Allard, R.W.: The mating system and microevolution. Genetics 79: 115-126. (1975).

Allard, R.W., Kahler, A.L.: Allozyme polymorphisms in plant populations. Stadler Symp. Vol. 3:9-24. Columbia, Mo. (1971).

Allard, R.W., Jain, S.K., Workman, P.L.: The genetics of inbreeding populations. Adv. Genet. 14:55-131. (1968).

Anderson, W.W.: Genetic equilibrium and population growth under density-regulated selection. Amer. Natur. 105:489-498. (1971).

Antonovics, J.: Evolution in closely adjacent populations. VI. Manifold effects of gene flow. Heredity 23:507-524. (1968).

Babbel, G.R., Wain, R.P.: Genetic structure of *Hordeum jubatum*. I. Outcrossing rates and heterozygosity levels. Can. J. Genet. Cytol. 19:143-152. (1977).

Bernstrom, P.: Cleisto- and chasmogamic seed-setting in di- and tetraploid *Lamium amplexicaule*. Hereditas 36:492-506. (1950).

Bocher, T.W.: Cytological and embryological studies in the amphiapomictic *Arabis holboellii* complex. Danske Videnskab. Selskab. Biol Skrifter 6:1-59. (1951).

Bodmer, W.F., Edwards. A.W.F.: Natural selection and the sex ratio. Ann. Human Genet. 24:239-244. (1960).

Bradshaw, A.D.: Some evolutionary consequences of being a plant. Evol. Biol. 5:25-47. (1972).

Bradshaw, A.D.: Environment and phenotypic plasticity. Brookhaven Symp. Biol. 25:75-94. (1974).

Breese, E.L.: Selection for differing degrees of outcrossing in *Nicotiana rustica*. Ann. Bot. 23:331-334. (1959).

Brown, W.V.: The relation of soil moisture to cleistogamy in *Stipa leucotricha*. Bot. Gaz. 113:438-444. (1952).

Bullock, S.H.: Consequences of limited seed dispersal within a simulated annual population. Oecologia 24:247-257. (1976).

Campbell, J.M., Abbott, R.J.: Variability of outcrossing frequency in *Senecio vulgaris* L. Heredity 36:267-274. (1976).

Charlesworth, B.: Selection in density-regulated populations. Ecology 52:469-475. (1971).

Charlesworth, B.: Selection in populations with overlapping generations. V. Natural selection and life histories. Amer. Natur. 107:303-311. (1973).
Charlesworth, B., Giesel, J.T.: Selection in populations with overlapping generations II. Relations between gene frequency and demographic variables. Amer. Natur. 106:388-401. (1972).
Christiansen, F.B.: Hard and soft selection in a subdivided population. Amer. Natur. 109:11-16. (1975).
Clark, R.B., Orton, E.R.: Sex ratio in *Ilex opaca*. Ait. Hort. Sci. 2:115. (1967).
Cohen, D.: Optimising reproduction in a randomly varying environment. J. Theoret. Biol. 12:119-129. (1966).
Cohen, D.: Optimising reproduction in a randomly varying environment when a correlation may exist between the conditions at the time a choice has to be made and the subsequent outcome. J. Theoret. Biol. 16:1-14. (1967).
Cohen, D.: A general model of optimal reproduction in a randomly varying environment. J. Ecol. 56:219-228. (1968).
Cohen, D.: Maximizing final yield when growth is limited by time or by limiting resources. J. Theoret. Biol. 31:299-307. (1971).
Cohen, D.: The optimal timing of reproduction. Amer. Natur. 110: 801-807. (1976).
Crow, J.F., Kimura, M.: An Introduction to Population Genetics Theory. Evanston, Ill.: Harper and Row, 1970.
Crowe, L.K.: The polygenic control of outbreeding in *Borago officinalis*. Heredity 27:111-118. (1971).
Demetrius, L.: Reproductive strategies and natural selection. Amer. Natur. 109:243-249. (1975).
Drayner, J.M.: Regulation of outbreeding in field beans. (*Vicia faba* L.). Nature 176:489-490. (1956).
Drayner, J.M.: Self- and cross-fertility in field beans (*Vicia faba* L.). J. Agric. Sci. 53:385-403. (1959).
Dzhaparidze, L.J.: Sex in plants. Pt-1. Akad. Nauk. Gruzinskoi SSR., Inst. Bot. (Trans. Israel Prog. Sci. Transl.). (1963).
Eshel, J.: Selection on sex ratio and the evolution of sex determination. Heredity 34:351-361. (1975).
Falconer, D.S.: Introduction to Quantitative Genetics. London: Oliver & Boyd, 1960.
Fisher, R.A.: The Genetical Theory of Natural Selection. Oxford: Clarendon Press, 1930.
Freeman, D.C., Klikoff, L.G., Harper, K.T.: Differential resource utilization by sexes of dioecious plants. Science 193:597-599. (1976).
Gadgil, M.: Dispersal: Population consequences and evolution. Ecology 52:253-261. (1971).
Gadgil, M., Bossert, W.: Life history consequences of natural selection. Amer. Natur. 104:1-24. (1970).
Gadgil, M., Solbrig, O.T.: The concept of r- and K- selection: evidence from wild flowers and some theoretical considerations. Amer. Natur. 106:14-31. (1972).
Ganders, F.R., Carey, K., Griffiths, A.J.F.: Outcrossing rates in natural populations of *Plectritis brachystemon* (Valerianaceae). Can. J. Bot. 55:2070-2074. (1977).
Gatina, Zh.L.: (Biological features of the sea buckthorn and the problem of its introduction into horticulture and forest belts). Problemy botaniki 2:339-374. lzd. An. SSR. (cited in Opler and Bawa, in press). (1955).
Giesel, J.R.: Fitness and polymorphism for net fecundity distribution in iteroparous populations. Amer. Natur. 108:321-331. (1974).

Gillespie, J.H.: Polymorphism in patchy environments. Amer. Natur. 108:145-151. (1974).

Gillespie, J.H.: The role of migration in the genetic structure of populations in temporally and spatially varying environments. I. Conditions for polymorphism. Amer. Natur. 109:127-135. (1975).

Gillespie, J.H.: The role of migration in the genetic structure of populations in temporally and spatially varying environments. II. Island models. Theoret. Pop. Biol. 10:227-238. (1976).

Gleaves, J.T.: Gene flow mediated by wind borne pollen. Heredity 31:355-366. (1973).

Godley, E.J.: Breeding systems in New Zealand plants. 3. Sex ratios in some natural populations. New Zealand J. Bot 2:205-212. (1964).

Grant, V.: The regulation of recombination in plants. Cold Spring Harb. Symp. Quant. Biol. 23:337-363. (1958).

Grant, V.: Plant Speciation. New York: Columbia Univ. Press, 1971.

Gustaffson, A.: Apomixis in higher plants. Lunds Universitets Arsskrift 39:1-370. (1946-1947).

Harding, J., Barnes, K.: Genetics of *Lupinus* X. Genetic variability, heterozygosity and outcrossing in colonial populations of *Lupinus succulentus*. Evolution 31:247-255. (1977).

Harding, J., Mankinen, C.B., Elliott, M.: Genetics of *Lupinus* VII. Outcrossing, autofertility, and variability in natural populations of the Nanus group. Taxon 23:729-738. (1974).

Harlan, J.R.: Cleistogamy and chasmogamy in *Bromus carinatus*. Amer. J. Bot. 32:66-72. (1945).

Harper, J.L.: Approaches to the study of plant competition. Symp. Soc. Exptl. Biol. 15:1-39. (1961).

Harper, J.L.: Establishment, aggression, and cohabitation in weedy species. In The Genetics of Colonizing Species. (ed. H.G. Baker and G.L. Stebbins). New York: Academic Press, 1965, pp. 243-265.

Harper, J.L.: Population Biology of Plants. New York: Academic Press, 1977.

Harper, J.L., White, J.: The dynamics of plant populations. Prov. Adv. Study Inst. Dynamics Numbers Popul. (Oosterbeek) 41-63. (1970).

Harper, J.L., White, J.: The demography of plants. Ann. Ecol. Syst. 5:419-463. (1974).

Harrington, J.F.: Seed storage and longevity. In Seed Biology Vol. 3. (ed. T.T. Kozlowski). New York: Academic Press, 1972, pp. 145-245.

Harris, W.: Environmental effects on the sex ratio of *Rumex acetosella* L. Proc. New Zealand Ecol. Soc. 15:51-54. (1968).

Haskell, G.: Adaptation and the breeding system in groundsel. Genetica 26:468-484. (1954).

Hedrick, P.W., Ginevan, M.E., Ewing, E.P.: Genetic polymorphism in heterogeneous environments. Ann. Rev. Ecol. Syst. 7:1-32. (1976).

Heslop-Harrison, J.: Photoperiod and fertility in *Rottboellia exaltata* L. Ann. Bot. 22:345-349. (1959).

Heslop-Harrison, J.: Sexuality of angiosperms. In Plant Physiology - A Treatise. (ed. F.C. Steward). New York: Academic Press, 1972, pp. 133-289.

Hjelmqvist, H., Grazi, F.: Studies on variation in embryo sac development. Bot. Not. 117:141-166. (1964).

Hjelmqvist, H., Grazi, F.: Studies on variation in embryo sac development. 2nd pt. Bot. Not. 118:329-360. (1965).

Holden, J.J.W., Bond, D.A.: Studies on the breeding system of the field bean, *Vicia faba* L. Heredity 15:175-192. (1960).

Holler, H.C., Abrahamson, W.G.: Seed and vegetative reproduction in relation to density in *Fragaria virginiana* (Rosaceae). Amer. J. Bot. 64:1003-1007. (1977).

Humphreys, M.O., Gale, J.S.: Variation in wild populations of *Papaver dubium*. VIII. The mating system. Heredity 33:33-41. (1974).

Il'in, A.M.: (Sex ratio in *Populus tremula* under different growing conditions.) Ekologiya 4:92-93. (1969). (cited in Opler and Bawa, in press.)

Jain, S.K.: Population structure and the effects of breeding system. In Crop Genetic Resources for Today and Tomorrow. (ed. O.H. Frankel and J.G. Hawkes). New York: Cambridge Univ. Press, 1975, pp. 15-36.

Jain, S.K.: The evolution of inbreeding in plants. Ann. Rev. Ecol. Syst. 7:469-495. (1976).

Jain, S.K., Bradshaw, A.D.: Evolutionary divergence among adjacent plant populations. I. Evidence and its theoretical analysis. Heredity 21:407-441. (1966).

Kahler, A.L., Clegg, M.T., Allard, R.W.: Evolutionary changes in the mating system of an experimental population of barley (*Hordeum vulgare* L.). Proc. Nat. Acad. Sci. 72:943-946. (1975).

Kaplan, S.H.: Seed production and sex ratio in anemophilous plants. Heredity 28:281-285. (1972).

Karlin, S.: Population subdivision and selection-migration interaction. In Proc. Int. Conf. Pop. Genet. Ecol. (ed. S. Karlin and E. Nevo). New York: Academic Press, 1976.

Karlin, S.: Gene frequency patterns in the Levene subdivided population model. Theoret. Pop. Biol. 11:356-385. (1977).

Karlin, S., Kenett, R.S.: Variable spatial selection with two stages of migrations and comparisons between different timings. Theoret. Pop. Biol. 11:386-409. (1977).

Karlin, S., McGregor, J.: The role of the Poisson progeny distribution in population genetic models. Math. Biosci. 2:11-17. (1968).

Kawano, S.: The productive and reproductive biology of flowering plants. II. The concept of life history strategy in plants. J. Coll. Liberal Arts, Toyama Univ. 8:51-86. (1975).

Kimura, M., Ohta, T.: Theoretical aspects of population genetics. Princeton, N.J.: Princeton Univ. Press, 1971.

King, C.E., Anderson, W.W.: Age-specific selection II. The interaction between r- and K- during population growth. Amer. Natur. 105:137-156. (1971).

Kivalaan, A., Bandurski, R.S.: The ninety-year period for Dr. Beals' seed viability experiment. Amer. J. Bot. 60:140-145. (1973).

Knox, R.B., Heslop-Harrison, J.: Experimental control of aposporous apomixis in a grass of the Andropogoneae. Bot. Not. 116:127-141. (1973).

Koller, D.: The physiology of dormancy and survival of plants in desert environments. Symp. Soc. Expt. Biol. 23:449-469. (1969).

Kolman, W.A.: The mechanism of natural selection for the sex ratio. Amer. Natur. 94:373-377. (1960).

Koyama, H., Kira, T.: Intraspecific competition among higher plants. VIII. Frequency distribution of individual weight as affected by the interaction between plants. J. Inst. Polytech. Osaka City Univ. (D) 7:73-94. (1956).

Leigh, E.G.: Sex ratio and differential mortality between sexes. Amer. Natur. 104:205-210. (1970).

Leverich, J.: Demographic studies of a population of *Phlox drummondii*. Ph.D. Thesis. University of Texas, Austin. (1977).

Levin, D.A.: Plant density, cleistogamy, and self-fertilization in natural populations of *Lithospermum caroliniense*. Amer. J. Bot. 59:71-77. (1972).

Levin, D.A., Kerster, H.W.: Neighborhood structure in plants under diverse reproductive methods. Amer. Natur. 105:345-354. (1971).
Levin, D.A., Kerster, H.W.: Gene flow in seed plants. Evol. Biol. 7:139-220. (1974).
Levin, D.A., Kerster, H.W.: The effect of gene dispersal in the dynamics of gene substitution in plants. Heredity 35:317-336. (1975).
Levins, R.: Evolution in Changing Environments. Princeton, N.J.: Princeton Univ. Press, 1968.
Lewis, D.: The evolution of sex in flowering plants. Biol. Rev. 17:46-67. (1942).
Lloyd, D.G.: Sex ratios in sexually dimorphic *Umbelliferae*. Heredity 31:239-249. (1973).
Lloyd, D.G.: Female predominant sex ratios in angiosperms. Heredity 32:35-44. (1974).
Lysova, N.V., Khizhnyak, N.I.: Sex differences in the trees in the dry steppe. Soviet J. Ecol. 6:522-527. (1975).
Madge, M.A.: Spermatogenesis and fertilization in the cleistogamous flowers of *Viola odorata* var. *praecox*. Hort. Ann. Bot. 43:545-577. (1929).
Marshall, D.R., Broué, P.: Outcrossing rates in Australian populations of subterranean clover. Aust. J. Agric. Res. 24:863-867. (1973).
Mayer, A.M., Poljakoff-Mayber, A.: The Germination of Seeds. London: Pergammon Press, 1975.
McKee, R., Hyland, H.L.: Apetalous and petaliferous flowers in *Lespedeza*. Agron. J. 33:811-815. (1941).
Mulcahy, D.L.: Optimal sex ratio in *Silene alba*. Heredity 22:411-423. (1967).
Nagylaki, T.: Dispersion-selection balance in localized plant populations. Heredity 37:59-67. (1976).
Nei, M., Imaizumi, Y.: Genetic structure of human populations. II. Differentiation of blood group gene frequencies among isolated human populations. Heredity 21:183-190. (1966).
Nei, M., Murata, M.: Effective population size when fertility is inherited. Genet. Res. 8:257-260. (1966).
Nygren, A.: Studies on vivipary in the genus *Deschampsia*. Hereditas 35:27-32. (1949).
Nygren, A.: Form and biotype formation in *Calamagrostis purpurea*. Hereditas 37:519-532. (1951).
Nygren, A.: Apomixis in angiosperms. II. Bot. Rev. 20:577-649. (1954).
Ogden, J.: The reproductive strategies of higher plants. II. The reproductive strategy of *Tussilago farfara*. J. Ecol. 62:291-324. (1974).
Opler, P.A., Bawa, K.S.: Sex ratios of tropical dioecious trees: Selective pressures and ecological fitness. Evolution, in press. (1977).
van der Pijl, L.: Principles of Dispersal in Higher Plants. New York: Springer-Verlag, 1969.
Putwain, P.D., Harper, J.L.: Studies in the dynamics of plant populations. V. Mechanisms governing the sex ratio in *Rumex acetosa* and *Rumex acetotella*. J. Ecol. 60:113-129. (1972).
Richards, A.J.: Notes on the sex and age of *Potentilla fruticosa* L. in upper Teasdale. Trans. of the Nat. Hist. Soc. Northumbria 42:84-92. (1975).
Rick, C.M., Fobes, J.F., Holle, M.: Genetic variation in *Lycopersicon pimpinellifolium*: Evidence of evolutionary change in mating systems. Plant Syst. Evol. 127:139-170. (1977).
Risser, P.G.: Competitive relationships among herbaceous grassland species. Bot. Rev. 35:251-284. (1969).

Roberts, E.H.: Viability of Seeds. London: Chapman & Hall, 1972.
Roberts, R.H., Struckmeyer, B.E.: Photoperiod, temperature and some hereditary responses in plants. J. Hered. 29:95-98. (1938).
Roff, D.A.: Population stability and the evolution of dispersal in a heterogeneous environment. Oecologia 19:217-237. (1975).
Ross, M.D., Harper, J.L.: Occupation of biological space during seedling establishment. J. Ecol. 60:77-88. (1972).
Sagar, G.R., Mortimer, A.M.: An approach to the study of the dynamics of plants with special reference to weeds. Appl. Biol. 1:1-47. (1977).
Salisbury, E.: A note on shade tolerance and vegetative propagation of woodland species. Proc. R.Soc. Lond. B. 192:257-258. (1976).
Saran, S., de Wet, J.M.J.: The mode of reproduction in *Dichanthium intermedium* (Gramineae). Bull. Torrey Bot. Club. 97:6-13. (1970).
Sarukhan, J.: Studies on plant demography. *Ranunculus repens* L., *R. bulbosus* L., and *R. acris*. II. Reproductive strategies and seed. J. Ecol. 62:151-177. (1974).
Sarukhan, J., Harper, J.L.: Studies on plant demography. *Ranunculus repens* L., *R. bulbosus* L., and *R. acris* L. I. Population flux and survivorship. J. Ecol. 61:675-716. (1973).
Schaffer, W.M.: The evolution of optimal reproductive strategies: The effects of age structure. Ecology 55:291-303. (1974a).
Schaffer, W.M.: Optimal reproductive effort in fluctuating environments. Amer. Natur. 108:783-790. (1974b).
Schaffer, W.M., Gadgil, M.D.: Selection for optimal life histories in plants. In Ecology and Evolution of Communities. (ed. M. Cody and J. Diamond). Cambridge, Massachusetts: Harvard Univ. Press, 1975, pp. 142-157.
Sheldon, J.C.: The behavior of seeds in soil. III. The influence of seed morphology and the behavior of seedlings on the establishment of plants from surface-lying seeds. J. Ecol. 62:47-66. (1974).
Spieth, P.T.: Theoretical considerations of unequal sex ratios. Amer. Natur. 108:837-849. (1974).
Stebbins, G.L.: Self-fertilization and population variability in higher plants. Amer. Natur. 91:337-354. (1957).
Stebbins, F.L.: Longevity, habitat, and release of genetic variability in higher plants. Cold Spring Harb. Symp. Quant. Biol. 23: 365-378. (1958).
Stone, D.E.: A unique balanced breeding system in the vernal pool mouse-tails. Evolution 13:151-174. (1959).
Suneson, C.: Frost-induced natural crossing in barley, and a corollary on stem rust persistence. Agron. J. 45:388-389. (1953).
Taylor, H.M., Gourby, R.S., Lawrence, C.E., Kaplan, R.S.: Natural selection of life history attributes: An analytical approach. Theoret. Pop. Biol. 5:104-122. (1974).
Thomas, A.G.: Productive strategies of *Hieracium*. Symp. on Plant Pop. Dynamics. AIBS Meetings, Tempe, Arizona. (1974).
Toole, E.H., Brown, E.: Final results of the Duvel buried seed experiment. J. Agric. Res. 72:201-210. (1946).
Vasek, F.C., Harding, J.: Outcrossing in natural populations. V. Analysis of outcrossing, inbreeding, and selection in *Clarkia exilis* and *Clarkia temblorensis*. Evolution 30:403-411. (1976).
White, J., Harper, J.L.: Correlated changes in plant size and numbers in plant populations. J. Ecol. 58:467-485. (1970).
Wilson, J.B., Levin, D.A.: Gene flow in plants in a patchy environment. Heredity, in press. (1979).
Wright, S.: Evolution in Mendelian populations. Genetics 16:97-159. (1931).

Wright, S.: Size of population and breeding structure in relation to evolution. Science 87:430-431. (1938).
Wright, S.: Isolation by distance under diverse systems of mating. Genetics 31:39-59. (1946).
Wycherley, P.R.: Vegetative proliferation of floral spiklets in British grasses. Ann. Bot. 18:199-227. (1954).
Yarranton, G.A., Morrison, R.G.: Spatial dynamics of a primary succession: Nucleation. J. Ecol. 62:417-428. (1974).
Younger, V.B.: Environmental control of the initiation of the inflorescence, reproductive structures, and proliferations in *Poa bulbosa*. Amer. J. Bot. 47:753-757. (1960).

6 Coda

Population Differentiation: Something New or More of the Same?

R.R. Sokal

A. Introduction

The general question addressed in this paper is a fundamental one for the synthetic theory of evolution. It concerns the assumption that the processes of evolution at the population level are sufficient to explain the diversity within and among species even at higher taxonomic levels. Simply put, is the evolutionary differentiation at intermediate and high systematic levels merely the accumulation of the results of microevolutionary processes over sufficient periods of time, or are there processes involved which are not ordinarily observed at the population level and whose mechanisms need to be elucidated? But since evolution takes place in a continuum of populations through time, all evolutionary processes must in the final analysis take place at the population level. The question then turns on whether the principles of population biology as propounded by the synthetic theory are sufficient to account for the types of evolutionary phenomena observed. In other words, does evolution beyond the demic level require something new, or is it more of the same?

Many mechanisms of evolution at the population level have been proposed, although there is considerable disagreement on the relative importance of the various processes. Comprehensive reviews and theoretical treatments of the effects and interactions of the various factors involved, e.g., mutation, recombination, chromosomal structure, migration, selection, and drift can be found in sources such as Wright (1969), Crow and Kimura (1970), Dobzhansky (1970), and Dobzhansky et al. (1977).

There are at least two distinct ways of approaching our problem. One is to survey evolutionary phenomena of various kinds and to look for those that cannot be explained by current evolutionary models. This has been the approach of those who, in turn, found the Darwinian thesis and, more recently, the synthetic theory of evolution insufficient to explain real or apparent phenomena, such as the origin of complex adaptations, the retention of rudimentary organs, or evolutionary trends such as anagenesis or orthogenesis. The overwhelming majority of biologists have maintained rather dogmatically for the last thirty years that the known principles of microevolution and speciation are sufficient to explain the variety of challenging evolutionary puzzles to which the skeptics keep referring (Dobzhansky, 1970; Mayr, 1963, 1970). Yet many evolutionists retain some doubts about the sufficiency of the synthetic theory, which itself is undergoing reexamination in the wake of the vigorous debates of recent years on selection versus neutrality, on group selection, and on the plausibility and occurrence of sympatric speciation. These problems have been reexamined in two recent important books. Frazzetta (1975) approaches the traditional questions with a refreshing lack of dogmatism but finds that the principles of modern population biology are sufficient to explain the phenomena considered. Riedl (1975, 1977) boldly postulates feedback from the epigenotype (the phenome) to the genome, compatible in his view with modern genetic theory and without violating the central dogma of genetics - that modifications to the phenotype do not affect the genotype. We shall clearly hear more about these and similar ideas in the years to come.

A second way of determining whether factors other than those apparent at the population level are required to explain some evolutionary phenomena is to undertake a systematic survey of variability and organization in diverse groups of organisms, to ascertain the nature and the amount of reorganization of the phenome and genome, and to determine the taxonomic level at which such reorganization is first manifested. Do the patterns of variation and covariation within demes first change among demes, geographic races, subspecies, species, genera, or even higher taxa?

Since it is generally impossible to watch populations or sets of populations evolve, the comparative and chorological approach applied throughout much of evolutionary biology is customarily employed. Populations are examined over space, their similarities and differences noted, and inferences made from these patterns to the nature of the processes that engendered them. So, specifically, the question posed is: do the patterns of variation observed among sets of populations within a species follow simply from the patterns of variation within the populations?

B. Variation

Variation may be viewed at several organizational levels. Immediately most obvious is genetic variation, which is usually measured nowadays as variation of enzyme polymorphisms. To what degree can genetic variability among populations be predicted as a function of intra-population variability?

I do not propose to review in depth the evidence on genetic inter-populational differentiation as a function of intrapopulational differentiation for two reasons. Although there are now numerous data sets available as a result of the massive electrophoretic efforts of the past ten years, there are few data sets designed to answer the questions we have in mind. Numerous local populations must be analyzed for a substantial number of loci, and for obvious reasons this has not been done. A second, more fundamental, difficulty concerns framing these questions precisely in genetic terms. How should average variation within local populations be quantified for a given locus? There is no agreement at the moment on a measure of intra-population variability for each locus which could be applied meaningfully to interpopulation variability as well. Many of the available measures such as average heterozygosity, average genetic divergence, Wright's F_{IS} and others, are necessarily related to measures of interpopulation variation, such as range of gene frequencies among populations, F_{ST}, or genetic identity.

A preliminary review of the relationships that have been observed can be limited to three studies. Using the data of Ayala, et al. (1974) Koehn and Eanes (1978) have recently shown that there is a correlation between intrapopulation heterozygosity and among-population genetic identity for monomeric and dimeric enzymes in the *Drosophila willistoni* group. The correlation is actually negative, since one of these variables is a similarity and the other one a distance measure. The authors conclude from these observations that there is a clear locus-specific contribution to among-population variation in genetic identity. Average within-population heterozygosity may be related to the molecular weight of the enzymes.

In a study of 22 populations of fishes from 5 species of the genus *Menidia*, Johnson and Mickevich (1977) were able to show a clear relationship between the information statistic within populations and interlocality allozymic diversity expressed as number of allelic substitutions. I have reanalyzed Schaal and Levin's data on *Liatris cylindracea* (Schaal, 1974, 1975; Schaal and Levin, 1976) and find no significant association between within population heterozygosity and either Wright's F_{ST} or interpopulation divergence as measured by range of gene frequencies. Thus from the evidence available to date, the relationship looks suggestive, but much more analysis of available data sets as well as new studies need to be done to confirm an association.

The question of whether interpopulation variation is merely an extension of intrapopulation variation can be answered at other organizational levels as well. Clearly, it needs to be asked at the level of the organization of the genome and in terms of chromosomal variation patterns. But here, too, the evidence is scant and insufficient at this time for a review.

I. Phenetic variation

Another way of viewing variation is in terms of phenotypes and phenetic variability. This is the approach I favor because of my own philosophical orientation and methodological bent. However, the phenetic approach to evolutionary dynamics, rather than the widely held genetic perspective, is also recommended by several other considerations:

Although the impression might not be obtained from classical genetic and electrophoretic studies, phenotypic variation is continuous, not discrete, in the vast majority of cases. The most common effect of a gene is quantitative rather than qualitative. In the study of variation, we are usually not interested in the qualitative differences among loci, but rather in differences among alleles at the same locus, which are typically dosage phenomena. There has been an historical bias in classical Mendelian genetics, as well as in allozyme genetics, towards factors with discrete effects. Since population geneticists tend to study one or a few loci, experimentalists have looked for cases where only a few loci are at work and where allelic substitutions at such loci have easily measurable effects. This rules out most characters of interest in evolutionary biology. Certainly characters involving fitness are generally quantitative. Even characters that appear discrete are often threshold effects with an underlying continuum produced by many genes, each furnishing a small contribution to the variance. Finally, alleles that are effectively discrete at the molecular level contribute to quantitative variation at a higher phenetic level. Milkman (1970) and others have pointed out that some variation at every level fails to be expressed at the next higher level of organization.

Despite the ever-increasing sophistication of population genetics, it is still quite inadequate for explaining changes in physiology, development, behavior and morphology, which are necessary to an understanding of many evolutionary phenomena. Furthermore, except in rather specialized instances, genotypes are tested at the organism-environment interface. This is, of course, the phenotype. Selection, therefore, acts on the phenome which is best viewed in a quantitative manner.

In view of the above one might argue that the problems of phenotypic variation should be approached through quantitative genetics -- the study of the inheritance of characteristics based on numerous small and additive effects. But after several decades of work, the state of quantitative genetics today is still insufficient to make strong predictions. It is usually possible to test whether any given variation is heritable, and to estimate heritability of a given character and the number of loci involved in a given trait. One can also predict the effects of directional selection, given a known distribution of phenotypes, an estimate of heritability, and specified directions and intensities of selection. However, this is insufficient for a theory of phenotypic evolution.

For these reasons it has seemed worthwhile to study phenetic variability in its own right, and in recent years various workers have turned to approaches that are, in essence, population phenetic. I need mention only the work of Lande, Roughgarden, Slatkin, Soulé, and Van Valen.

To examine the problem of whether differentiation among populations is merely the result of continuing processes within populations, a first look might be at variances of characters. In typical Mendelian populations, where each individual represents a different diploid genotype, the variance of a phenotypic character may represent either genetic variability or phenotypic variation as a result of environmental and developmental variability, or most likely, both phenomena. The distribution of a phenotypic variable thus results from the distributions and magnitudes of the underlying genetic and environmental factors. A phenotypic distribution may reflect selection pressures of the immediate past. A recent striking example of this has been furnished by Kerfoot (1975) studying a polymorphism in the cladoceran *Bosmina longirostris* in a bay in Seattle, Washington. Apparently in response to strong visual predation by fish, the inshore population possesses short mucrones and antennules. By contrast, these organs are long in the offshore populations, apparently as a defense mechanism against the grasping predatory copepods *Epischura*. These population differences are produced seasonally as the *Bosmina* populations and their predators increase after a winter low.

Phenotypic distributions are also believed to reflect the distribution of selective forces through evolutionary history, but while undoubtedly true in general, this is quite difficult to demonstrate in a specific case. Does the distribution of beak sizes in a bird population match exactly that of sizes of seeds or insects that it feeds on? Does the distribution of duration of larval period in a fly population reflect that of temporal food availability? Although causal links between such distributions are frequently assumed, e.g., Roughgarden (1972), they are by their nature very difficult to prove.

II. Variability profiles

To compare the amount of variation of different structures we employ the well known coefficient of variation, a measure of the standard deviation as a proportion of the mean. One way to represent the coefficient of variation for a large number of characters in a population is by means of a so-called variability profile (introduced by Yablokov, 1974). This is a graph of the coefficients of variation of a suite of characters in a population arrayed in some logical sequence. Usually, but not necessarily, these characters are morphological. Variability profiles point out characters of especially high or low variation and enable contrasts to be made between suites of equitably and differentially varying characters or among suites of characters from different populations of the same or of different species. Figure 1a illustrates a variability profile of 31 characters in the

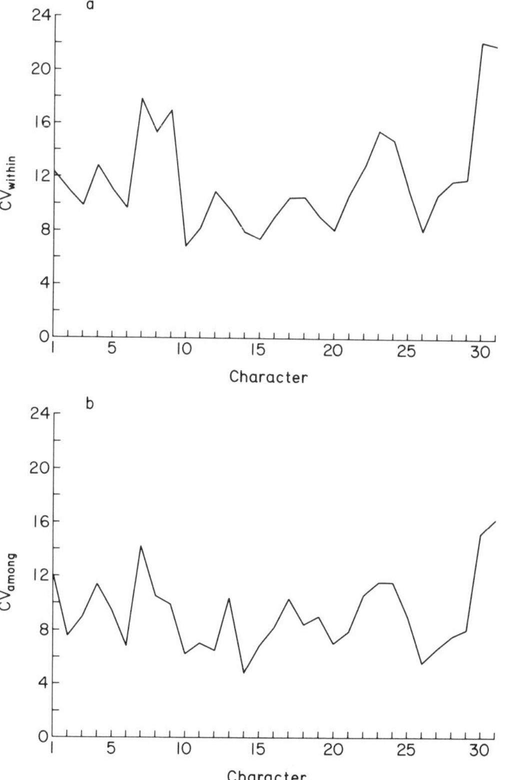

Fig. 1. Variability profiles within and among localities for stem mother and alate characters of the aphid *Pemphigus populicaulis*. The stem mothers are based on 118 localities; the alates on 101 localities. Along the abscissa the characters are arrayed in conventional order while the ordinate indicates coefficients of variation. The numbers along the abscissa refer to the characters. Numbers 1-9 are stem mother characters; 10-31 are alate characters.
a) Variability profile based on pooled variances within localities.
b) Variability profile based on variance components among localities ($CV = 100\ \sigma_A/\bar{Y}$).

North American aphid species, *Pemphigus populicaulis*, based on pooled variances within 108 local population samples over virtually its entire geographic range (Sokal and Bird, in preparation). Substantial differences in coefficients of variation are shown by these characters of stem mothers and alates with values ranging from 7 to 22.

Why would characters differ in their variabilities? As a general rule, they differ because of selective constraints operating at different amplitudes in different regions of the character suite. These constraints may be direct, as in selective elimination of unfit individuals by stabilizing selection. They also may be indirect, in that variability of a given character is controlled by developmental forces which determine the variation through morphogenetic correlations relating the variable in question to another variable actively determined by the environment. It must not be implied that stabilizing selection is the only way in which natural selection can bring about differential variability in characters. Obviously, directional and disruptive selection will both affect character variability, as numerous authors have pointed out (for reviews see Yablokov, 1974; Wright, 1977).

When a series of populations is studied and the same characters are analyzed in each population sample, variances can be computed within and among populations. Such a design possesses the structure of single classification analysis of variance. The expected value of the variance of the sample means would be σ^2/n_0 where σ^2 is the parametric variance of the character within populations, estimated by MS_{within}, the pooled error mean square of the anova; and n_0 is an average sample size. Should there be added differentiation among localities the variance among means would represent $\sigma^2/n_0 + \sigma^2_A$, where σ^2_A is the added variance component among localities.

Unless variability profiles are uniform over the range of populations studied, it is difficult to generalize about the extensions of intrapopulation variabilities to interpopulation variabilities. But if, for example, character X is considerably more variable than character Y in <u>all</u> populations, then it becomes of interest to know whether there is more added differentiation among populations for the more variable character X than for Y. It has been the experience of persons who have investigated variability profiles (e.g. Yablokov, 1974) that for many local populations of a species the variability profiles remain quite constant. That observation, made mainly on mammals by Yablokov, is being extended in my laboratory to observations on local population samples over the entire range of several aphid species. The work is still in progress, but so far it appears that variability profiles are reasonably homogeneous.

Identity in variability profiles of populations sampled over a wide range implies that the distribution of environmental or developmental constraints is constant across the localities inhabited by these populations. It also implies that the niche configurations for these populations are similar. (Of course, this means only configurations for those portions of the hyperdimensional space reflected in the phenome under investigation; configuration here means the shape of the realized niches for the populations being considered. It does not mean dimensionality of the fundamental niche, because that is presumably similar for all populations whether the variability profiles are parallel or not. If the variability profiles are parallel then the relevant regions of the niche space should be similar.)

Variation in variability profiles among populations may take several forms. The variability profiles may be parallel but differ in height, indicating that some populations are more variable than others but proportionately so for all characters considered. In suites of largely uncorrelated characters, I would predict that such a situation would rarely be found, because of the cost of selection argument. It is improbable that one could find populations in which numerous functionally independent characters are under simultaneous selection. Changes in "overall variability" (presumably for all characters) have been postulated by various authors. Such changes have been hypothesized for peripheral populations versus central ones, for ecologically marginal situations, and for areas subjected to large amounts of gene flow. The evidence on this subject is contradictory so far (Soulé 1967, 1972; Soulé and Baker, 1968; Senner, 1977). Populations of lizards, butterflies and aphids do appear to differ in overall variability, but in the most detailed analysis of such data to date, Senner was unable to relate differences in variability to any of the hypothesized causes.

Nonparallel variability profiles over the range of population samples studied imply both a raising or lowering of the level of variability as well as a change in profile. The variability of some characters will thus increase or decrease with respect to the variability of other characters along the variability profile. Such changes imply variation in the configurations of the niche subspaces, i.e., realized niche widths may be wide in one population but narrow in another population. The changes may also necessitate the making or breaking of character correlations resulting in new profiles.

III. The Kluge-Kerfoot phenomenon

On the assumption of homogeneity of variability profiles over local populations one may pool the variances within populations for each character. For all characters one can then plot some measure of among-population variation against the pooled within-population variances, all converted to coefficients of variation to make them comparable. For the within-population variation measure one may also compute a weighted mean CV within populations. Such scattergrams were first suggested by Kluge and Kerfoot (1973) who reported marked correlation between inter- and intra-population variability for numerous characters in several vertebrate species. In a recent paper (Sokal, 1976), I have reanalyzed their data sets and some additional ones by plotting coefficients of variation based on the square root of the variance component, $s_A/\bar{\bar{Y}}$, against CV within populations, $s/\bar{\bar{Y}}$, for each character and calculating Kendall's rank correlation τ to test significance of a scattergram. The findings still hold, although less strongly than suggested by these authors: characters that are more variable within populations are differentiated more among populations in terms of added variance components, independent of pooled within-group variances.

More recently I have analyzed two new data sets. The first is on humans, from the recent, extensive study of human variability in 18 Melanesian villages on the island of Bougainville (Friedlaender, 1975; Rhoads and Friedlaender, 1975). The data analyzed are 13 morphometric variables of the body and head of adult males. There is considerable differentiation in morphometry (as well as gene frequencies and language) among the villages despite their relative geographic proximity. For the 13 characters a significant rank correlation ($\tau = 0.564$; $P < 0.05$) exists between the coefficient of variation among and within villages (see Figure 2a). The other analysis is based on the data by Sokal and Bird (in preparation) for the aphid *Pemphigus populicaulis*, whose variability profile is shown in Figure 1. The coefficients of variation among 31 morphometric characters plotted against the coefficients of variation within the locality samples shown in Figure 2b yield a significant rank correlation coefficient of $\tau = 0.609$ ($P < .001$).

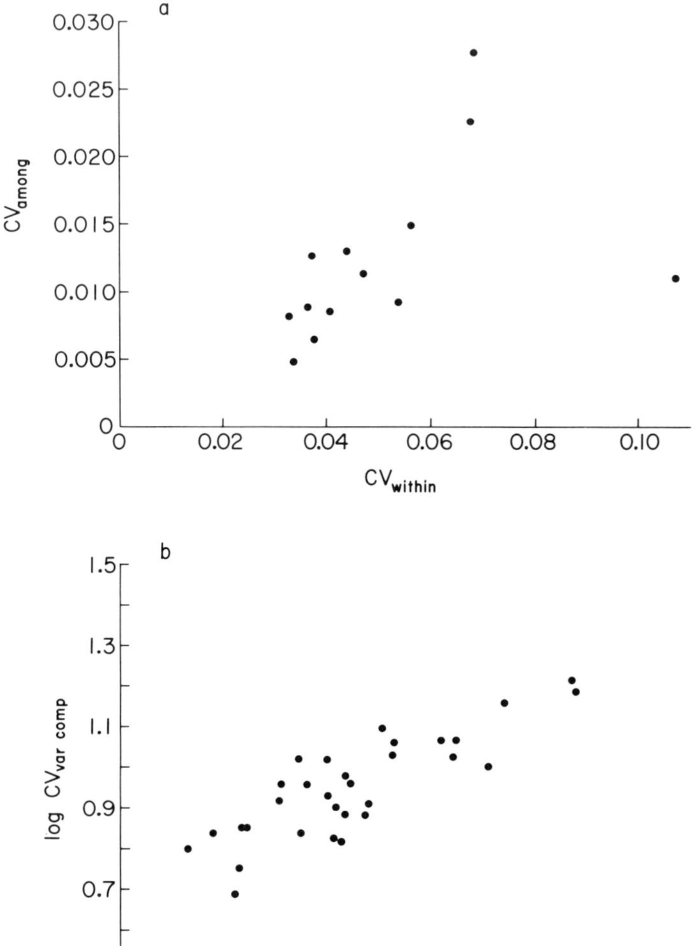

Fig. 2. Bivariate plots of coefficient of variation based on variance component among localities (ordinate) against coefficient of variation within localities (abscissa). a) Morphometric variables in human populations from Bougainville (from Friedlaender, 1975). b) Morphometric variables in populations of the aphid *Pemphigus populicaulis* (from Sokal and Bird, in preparation). Details on numbers of localities and characters can be found in the text.

The Kluge-Kerfoot phenomenon is surprising in several respects. It seems rather unexpected that diversification among populations and over large geographic areas should be along the same dimensions as variability within populations. Does this imply that the variability profiles have been stable for substantial periods of evolutionary time and that population differentiation is merely an extension of the trends in these populations? This is a kind of uniformitarianism in evolutionary patterns which is of considerable interest. Do selective agents affecting variability within populations vary along the same dimensions among populations? This is conceivable, for example, in a situation where a size distribution of a resource for a population A is similar to that of a population B, except that for population B it is shifted in some direction. But given the probable complexity of the niches of populations, it seems unlikely that the selective constraints differentiating populations within a species will be identical to those determining variability within populations. I would therefore expect patterns of among-population variability not to be too closely related to within-population variability. Yet when we compare such patterns we do not find substantial differences. Such comparisons can be made by means of Kluge-Kerfoot type scattergrams as shown in Figure 2. However, a clearer impression is gained when the actual variability profiles are compared. Figure 1 shows an among-population variability profile (Figure 1b) and a within-population profile in (Figure 1a). Although the coefficients of variation within populations are generally higher than those based on variance components among localities, the peaks and troughs of the two profiles generally correspond quite well.

If the environmental spectrum among populations is indeed different in direction and magnitude from that within populations, the existence of the Kluge-Kerfoot phenomenon may be due to constraints of the developmental systems giving rise to these characters, rather than to the homology of selective constraints within and among populations. Kerfoot, in Kluge and Kerfoot (1973), has also suggested that "intrinsic" regulation based on functional relationships between characters may be responsible for determining levels of variance for characters in a local population. In a similar manner the differences in character means among populations may lead to interpopulational differential variability because of such functional relationships.

The model considered so far envisages interpopulation variability as a function of intrapopulation variability. However, the reverse model is at least theoretically possible. A Kluge-Kerfoot effect can be generated from populations that had differentiated for various characters without regard to the within-population variability. Subsequent migration with or without interbreeding would result in increases in the within-population variance in proportion to the interlocality divergence of each character. Only under a particular combination of historical circumstance and population structure would such a model be plausible.

If the Kluge-Kerfoot effect is strongly present this is *prima facie* evidence that the variabilities involved have a genetic base. It would be unlikely that developmental noise of all characters varies in such a way that those characters with the largest noise within localities differed most among localities. Even though a model can be constructed where a character, such as size, might be directly affected by nutritional differences, it is improbable that the variation of an entire suite of characters would be entirely determined by environmental variation both within and among populations. This prediction is not fulfilled in the following data sets. In thirteen characters of the ciliate *Plagiopyliella pacifica*, an entocommensal of sea urchins studied by Lynn and Berger (1972), I found a high value of τ = 0.692 for the correlation within and among monodemes, populations obtained from individual urchins (Sokal, 1976). Berger (1978) believes that these monodemes differ genetically from each other but that there is no evidence for sexual reproduction within monodemes. These, then may represent clones, making character variation within monodemes nongenetic. The strong Kluge-Kerfoot phenomenon in this data set would lead me to suspect that the monodemes are in fact genetically variable. In a recently analyzed study of monodemes of the ciliate *Ancistrum mytili*, entocommensal in mussels of the genus *Modiolus*, Berger (personal communication) found an even stronger Kluge-Kerfoot effect (τ = 0.887), the highest such value found in any data set so far. However, in this ciliate monodemes are known to be genetically diverse. Thus, while a Kluge-Kerfoot effect cannot be ruled out for cases in which variation at both levels is entirely environmental, it seems unlikely on theoretical grounds.

When the total information currently available on Kluge-Kerfoot type analyses is summarized by plotting τ, the measure of rank correlation between inter- and intrapopulation coefficients of variation, against the taxonomic rank of the populations involved in the comparison, a potentially interesting relationship may be emerging (Figure 3). The

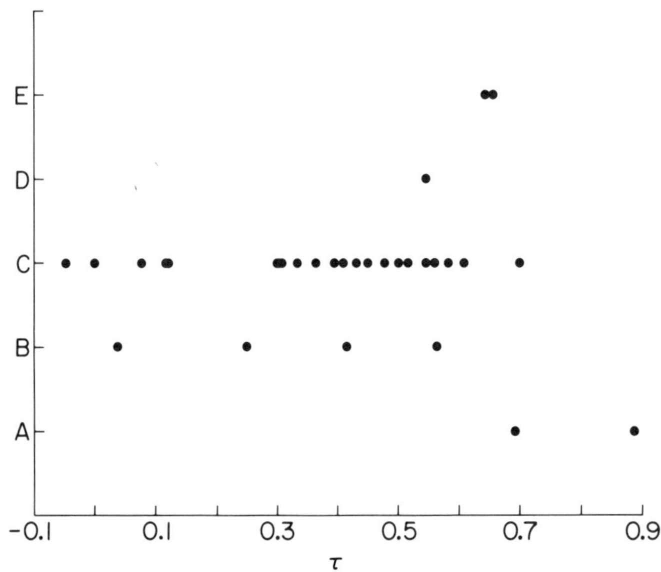

Fig. 3. Relation between measure of Kluge-Kerfoot phenomenon and taxonomic rank of populations involved in the comparison in 31 studies. Abscissa: τ, the rank correlation between inter- and intrapopulation coefficients of variation. Ordinate: Rank of taxonomic groups being tested for the Kluge-Kerfoot phenomenon. A -- Monodemes among host individuals. B -- Populations among localities in a microgeographic setting. C -- Populations among localities from the range of a species. D -- Populations from a species complex. E -- Populations from several species within a genus. The data points plotted comprise all the analyses published by Sokal (1976), the two new values first reported in this paper, and the value for *Ancistrum* reported by Berger (personal communication).

highest average values of τ occur at the lowest and highest hierarchic ranks at which such tests have been carried out -- among hosts of endosymbiotic ciliates and among populations of species within a genus in two vertebrates. Kluge and Kerfoot (1973) make the point that widely ranging populations will be subject to a wider range of environmental factors and should exhibit the Kluge-Kerfoot phenomenon more clearly. Except for the ciliate data, their prediction is borne out so far. For microgeographic studies a range from zero to moderate correlations is recorded, whereas local populations within a species range widely from very high to negligible values. Is the restriction of high Kluge-Kerfoot effects to the two extremes in taxonomic rank a real phenomenon or is it an artifact due to the relatively arbitrary allocation to categorical ranks, which depends on the extent of phenetic differentiation and the observer's perception of the population structure of the organisms studied? While it may, of course, be that in a population biological sense the ciliate monodemes studied are more differentiated among individual hosts than local populations of sparrows, aphids, or houseflies across a continental range,but such a possibility seems doubtful. However, if on further accumulation of Kluge-Kerfoot type scattergrams for other organisms the relations suggested here are borne out, it may lead to interesting reinterpretations of the phenomenon.

IV. Population Differentiation

In thinking about the Kluge-Kerfoot phenomenon one might consider how patterns of population differentiation may have arisen and whether different modes of origin give rise to different relations between variation among and within populations. At least four major models might be distinguished. In the first a character is differentiated in response to an environmental gradient producing a cline. In the second, environmental patches are heterogeneous among patches but homogeneous within. The heterogeneity among patches may or may not exhibit a pattern. A third proposes differentiation due to the classical isolation-by-distance model. A fourth ascribes the differentiation to historical factors of several kinds. Accidental immigration and the establishment of different types (the so-called founder effect resulting in a patchy distribution) or accidental settlement by two or more types with subsequent production of gradients between them by diffusion processes come to mind readily. Another historical factor may involve migration into an area by several populations that had differentiated elsewhere, which may be a more common phenomenon than is recognized. Human cultural analogs come to mind readily. The pattern of dialectical differentiation in the United States reflects earlier linguistic differentiation by different immigrant groups from the British Isles and other regions. Such cases, of course, only complicate the analysis of the origin of the differentiation. In actuality none of these models is likely to provide the sole explanation for any one character. Real situations in nature will more likely involve mixtures of two or more models.

In the first two models selection brings about population differentiation. Given the same selection intensity among populations for all characters, the interpopulation differentiation for each character should be proportional to its heritability. Thus, if the differentiation were simply a function of evolutionary time and all characters had the same heritability, one should expect a straight-line Kluge-Kerfoot scattergram. However, a straight-line Kluge Kerfoot relation is not restricted to the above case, as can be seem by examining the familiar equation $\underline{R} = \underline{i}\ \sigma\ \underline{h}^2$ (Falconer 1960), where $\underline{R}$ is the response to selection, $\underline{i}$ is the selection intensity in standard deviation units, σ is the standard deviation of the phenotypic character and $\underline{h}^2$ is the heritability. So long as $\underline{ih}^2$ is constant for all characters, a straight-line Kluge-Kerfoot relationship will result. Since there is no *a priori* reason for either selection intensities or heritabilities of different characters to be the same in a set of populations, there is no reason to expect a straight-line relationship. In fact no such example has come to light. The fact that there is at least a tendency for the two CV's to be directly related in most cases suggests that the selection intensities and heritabilities of characters investigated in these studies vary between closer bounds than might be anticipated.

The last two models imply stochastic differentiation among the populations. Regardless of whether differential variability is based on differences in inherent variability or in selective constraints, characters with a narrow variation range can be expected to be less variable among populations. This is because relatively few deviant founder populations are produced which might reach new adaptive peaks at different locations. By contrast, characters with broad variation ranges will more frequently give rise to phenodeviant isolates and can probably lead to widely different populations at other localities.

The above models of population differentiation are not immune from the all-pervading controversy between the neutralists and the selectionists. The increased interpopulation differentiation for those characters that are most variable within populations is compatible with both a selectionist hypothesis and a random-genetic-drift, isolation-by-distance model. Is there no hope for telling these two apart? In the case of population differentiation, the technique of spatial autocorrelation developed by Cliff and Ord (1973) and adapted for population biology by Sokal and Oden (1978a,b), may be able to distinguish these situations. This technique tests whether the observed value of a nominal, ordinal, or interval variable at one locality is independent of values of the variable at neighboring localities. The method was extended by Sokal and Oden (1978a) to include the computation of spatial correlograms which show changes in the autocorrelation coefficients as a function of increasing distances in the study area and summarize the patterns of geographic variation exhibited by the response surface for any given variable. Thus they are a simple kind of spectral analysis for response surfaces.

Spatial correlograms describe the underlying spatial relationships of a surface, and for this reason they are probably closer guides to some of the processes that have generated the surface than are the surfaces themselves. While identical variation patterns will yield identical correlograms, different variation patterns may or may not yield different correlograms (Sokal and Oden, 1978b). Conversely, different correlograms must be based on different patterns, but identical correlograms may result from identical or different patterns. The response surface observed is presumably a result of both fixed underlying influences (environmental patches, etc.) and influences that, by their very nature, cause autocorrelation (migration, etc.). If two variables have different patterns but similar correlograms, this probably means that both variables are largely determined by the same autocorrelated process (e.g., migration). If, however, the patterns are the same, this probably means that exogenous, fixed influences, are significant determinants of the surface.

When different characters (or gene frequencies) for the same populations in a given area show the same variation pattern and therefore identical (or similar) correlograms, this suggests a common response to selective agents that vary over the area. Other explanatory hypotheses for such correlations, such as several selective agents themselves correlated over space to which different characters respond in a necessarily correlated manner, are less probable. Morphometric characters also could be correlated over space because of pleiotropy.

Different variation patterns yielding the same correlograms should occur in populations where migration strongly affects geographic variation of gene frequencies or morphometric characters. The several gene frequency patterns need not correspond for historical reasons, such as different immigration waves or because of differential selection patterns. But since a migrant individual must carry with it all its genes, autocorrelation should be the same for all loci, and correlograms, or at least portions thereof, should resemble each other. Similar spatial autocorrelation of separate environmental variables leading to different variation patterns but similar correlograms is conceivable but unlikely.

Different geographic variation patterns associated with different correlograms should be found in populations in which differential selection among loci, rather than migration, predominates. This should be the most frequent pairing of conditions encountered in nature. The higher order autocorrelations could also differ for historical reasons such as invasion by populations differing for one character but not the other. In this discussion, I have ignored the possibility of gene frequency dependent migration which would further complicate the conclusions.

An example of such an approach is found in the analysis by Sokal and Oden (1978a) of the work of Selander and Kaufman (1975) describing microgeographic variation in allozyme frequencies in the European brown snail *Helix aspersa*. Although this example is on a microgeographic basis with no evidence of a Kluge-Kerfoot type effect, the analysis of this data set illustrates the type of reasoning that would go into distinguishing selective from stochastic patterns of population differentiation. The snails were collected exhaustively from shrubbery in two adjacent city blocks in Bryan, Texas. The allozyme variation was analyzed for 5 polymorphic enzymes, and the gene frequencies of each population were mapped. Although Selander and Kaufman showed both blocks to be significantly heterogeneous for most loci, these authors found the two blocks to differ considerably in pattern when tested by the method of Royaltey, Astrachan and Sokal (1975). Of seven patterns in Block A only one was significant, while seven out of ten of the allozyme patterns in Block B were found significant by this test. Figure 4 illustrates the microgeographic variation of two allozyme frequencies from Block A, both significantly heterogeneous among localities but lacking significant spatial patterns. These can be contrasted with the similarity of significant patterns in Block B. The correlograms obtained by spatial autocorrelation analysis of all allozyme frequencies are shown in Figure 5. Block A has only 1 significant positive autocorrelation over distances of 10 meters and few negative autocorrelations at greater distances. No trend in autocorrelations with increasing distance could be found. By contrast most allozyme frequencies in Block B show a consistent pattern of significant positive autocorrelations at 10 meters and of significant negative autocorrelations at distances from 50 to 90 meters. Thus autocorrelation in Block B tends to decrease with increasing distance. Neighboring populations in Block B are more

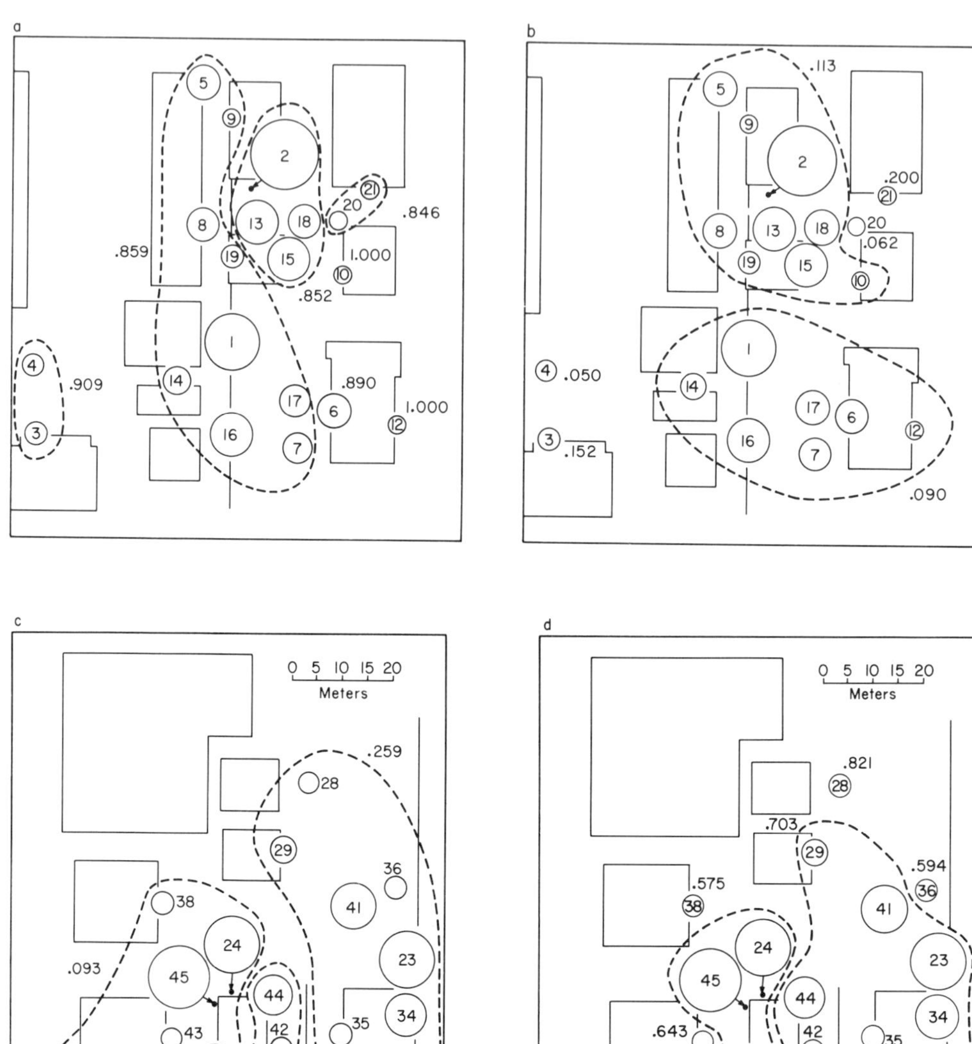

Fig. 4. Allele frequencies in colonies of the snail *Helix aspersa* in two city blocks in Bryan, Texas. Data from Selander and Kaufman (1975). a) Lap 1 allele 1 in Block A. b) Lap 2 allele 4 in Block A. c) Lap 1 allele 1 in Block B, d) PGM-1 in Block B. Colonies are indicated by circles, whose sizes are proportional to colony size. Numbers within circles are colony code numbers. The heavy lines enclose statistically homogeneous (maximally acceptable) and geographically connected localities defined by the method of Sokal and Riska (in preparation). Proportions next to a colony or an enclosed set of colonies indicate mean allele frequency.

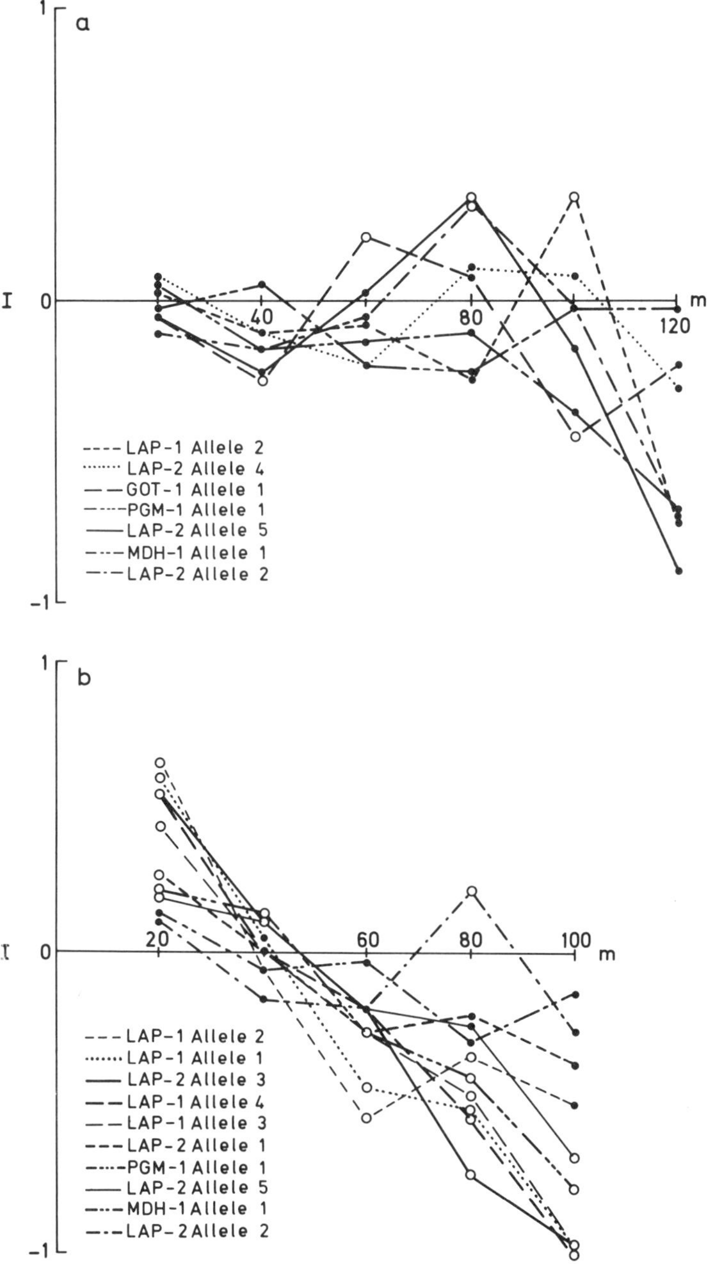

Fig. 5. Correlograms for geographic variation patterns of allozyme frequencies of the snail *Helix aspersa* in two blocks in Bryan, Texas a) Block A. b) Block B. Abscissa: shows distance in meters. Distances indicated are upper limits of the distance class rather than class marks. Ordinate: Moran's autocorrelation coefficient I. Significant autocorrelation coefficients are indicated by open circles. Based on data from Selander and Kaufman (1975).

similar to each other than are such populations in Block A.

In Block A the heterogeneous allozyme frequencies show little significant spatial structure, and what little exists is not similar for various allozyme frequencies. Because there is not much pattern, a stochastic interpretation is favored. Perhaps several immigration events led to founder effects. A selectionist interpretation -- i.e., different sites differ in various selective agents that affect the various allozyme frequencies--is improbable in view of the parallel variation for these same frequencies in Block B. That block looks very different. All significantly heterogeneous allozyme frequencies in this block have similar and significant patterns as shown by the Royaltey, Astrachan and Sokal test (Royaltey et al., 1975) and by the correlograms (compare Figures 5a and 5b). One might therefore suppose that similar selection had occurred. In view of the diversity of loci studied, this seems unlikely. It would be more probable to assume a migratory event followed by diffusion, giving rise to a situation mimicking an adaptive cline. This would also explain the discrepancy between the correlogram patterns for Block A and Block B. If these loci are subject to strict selective control, one would have to postulate important differences in the microclimatic and microenvironmental substructuring of both blocks. Selander and Kaufman offer no basis for this conjecture. An alternative explanation for the differences between the blocks might be that there were substantial differences in time since colonization. If Block A had been colonized early, but Block B more recently, the pattern in A could be interpreted as selection due to diverse microhabitats, whereas that for B could be diffusion between two diverse immigrant stocks, before selection had patterned the samples to the microhabitats in B. However, we have no evidence for such a sequence of events.

C. Covariation

Similarity in variability profiles of numerous populations of one species indicates the likelihood, but not the necessity, that covariation of characters within populations over numerous localities remains constant as well. This might be expected since covariation within localities may express the developmental coadaptations of the phenome of the species population and is unlikely to change over relatively short spans of space or time. But evidence from the Kluge-Kerfoot scattergrams indicates some nonparallelism between intralocality and interlocality variability profiles. This suggests that interlocality correlations may also differ somewhat from intralocality correlations. The immediate genetic and developmental causes may be the same in inter- as in intralocality correlations: linkage between genes or gene complexes, pleiotropic actions of the same gene complex, forced correlation of the body parts through selection, or common sensitivity to an environmental factor. Yet the interlocality correlation may differ from the intralocality correlation in magnitude and direction, since the action of these factors over the area of distribution of a species should frequently be unlike that within single populations.

The relation between intra- and interpopulation covariation is therefore of interest empirically and theoretically. First let us visualize the development of character correlations within populations. Figure 6a is an idealized drawing of the circular isodensity contour of two uncorrelated, normally distributed characters. Since the characters under discussion are quantitative, and assuming additive genetic variance for both characters, one may postulate a 1 - α isodensity contour or isophene for the population, shown by the outer circle. One may also superimpose circular isofitness contours onto the two-dimensional phenetic space. These contours, shown in Figure 6b, decrease uniformly as distance from the bivariate mean increases in any direction. Now let us apply normalizing selection. In the familiar univariate case the distribution of fitnesses along the phenetic axis would contract, and selection would eat into both tails of the distribution of phenotypes, maintaining the mean but reducing the variance of the population. In the bivariate case normalizing selection implies reduction of the values of the circular isofitness contours, and results in a population with equal bivariate mean and smaller dispersion. This is shown by the dashed concentric circle of smaller radius in Figure 6a.

Correlation implies distortion of the circular isofitness contours. Not all phenotypes on a given circular isodensity contour line will be equally fit, resulting in differential selection. The fitness contours are now ellipsoidal (Figure 6c) and represent the surface of an elliptic mound. In Figure 6d we show the result of such correlating selection after one or more generations, while maintaining the variance of each character. The arrows indicate the approximate direction of selection. Whether actual directions are the geodesics across the fitness surface or follow more complicated trajectories will depend on the genetic architecture of the characters involved (Lewontin, 1974).

We may expect that in the generations following selection, the Mendelian process will reconstitute the original dispersion. But repeated selection in case 6a and b with *concentric* equal fitness contours will reduce the variances of the two characters and lead to a reduced dispersion in which any given isodensity contour, say the 95% contour, now has a shorter radius in phenetic space. This is simply canalization in two dimensions. In cases 6c and d where isofitness lines are not equidistant from the bivariate mean, repeated selection will give rise to phenetic linkage of the characters through one of the genetic or developmental mechanisms mentioned earlier. Instead of a circular phenetic distribution of offspring, future generations will produce an ellipsoidal one, saving considerable wastage of unfit phenotypes. Occasional deviant phenotypes produced in the off-quadrants of the graph will be eliminated promptly.

The fitness surface in Figure 6c has been drawn to impose correlation while maintaining the standard deviations of both variables. The model need not be so restricted. The adaptive surface underlying Figure 6e produces an ellipse whose major axis is shorter than the diameter of the original circular distribution of the variables. Selection here is correlating and normalizing. All three models result in canalization of the variables in two-dimensional phenetic space.

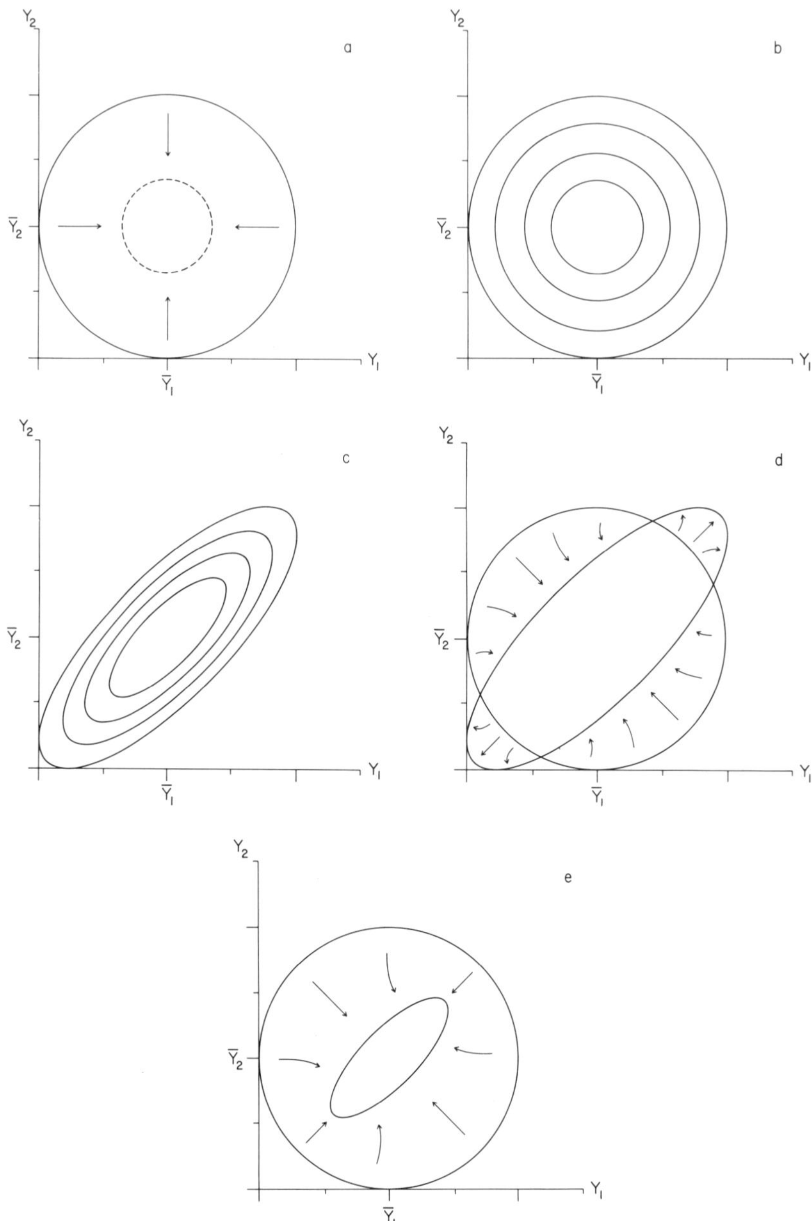

Fig. 6. Schematic diagram to illustrate development of character correlations. a. Y_1 and Y_2 are phenetic axes representing two characters. The outer (solid) circle is the $1 - \alpha$ isophene of the joint distribution of the two characters when they are normally distributed but uncorrelated. The inner (dashed) circle illustrates the reduced dispersion of the two characters after bivariate normalizing selection based on the isofitness contours of Figure 6b was applied. The arrows indicate the direction of selection. b. Isofitness contours drawn into the phenetic space shown in Figure 6a.

Fitnesses of phenotypes decrease with distance from the center of the circles. Fitness is highest at the bivariate mean. Thus these contours describe a bell-shaped mound and lead to bivariate normalizing selection. c. Isofitness contours drawn into the phenetic space shown in Figure 6a. Fitnesses of phenotypes decrease with distance from the center of the ellipses. Fitness is highest at the bivariate mean and decreases more rapidly along the minor axis of the ellipse than along the major axis. These contours describe an elliptic mound and lead to correlating selection. d. Results of correlating selection. The isophenes shown in Figure 6a are modified by the correlating selection imposed by the fitness contours of Figure 6c. The circular distribution of phenotypes is changed into an elliptical one with identical variance. Arrows indicate the direction of selection. e. Results of correlating and normalizing selection. Isofitness contours (not shown) are such as to produce correlated characters with reduced variances.

Character correlations among localities may have different origins depending on whether and how much those same characters are correlated within localities. If there is correlation within localities (i.e. among the phenotypes of a local population) the phenetic response to shifting environmental factors will be easiest along the principal axis of the ellipse of phenotypes, which is the axis of maximal phenetic displacement. In effect the ellipsoidal isofitness contours would be translated along this axis. In this model the various local populations subjected to different environments would continue to have a common slope and the identical Y-intercept, but changed means. Such a situation, illustrated in Figure 7a, would result in tight interlocality correlation. Remember that while the several evolved populations can be shown in a single phenetic space as in Figure 7a or 7b, the fitness surface would change for each locality and would in principle require a separate graph. In Figure 7b the local populations occupy phenetic space so as to exhibit no interlocality correlation although intralocality correlation is as strong as before. While the slopes of the local populations continue parallel, the Y-intercepts and means of various populations have been shifted. Such changes, if they occur, cannot be easy to accomplish. Translation of the original isodensity ellipse in the phenetic space into the several ellipses shown requires distortion in the fitness surface to admit selection of new forms. Such changes should involve complicated rearrangements of the genome of the populations and are probably rare. A situation in which some translation of the Y-intercept is permitted by the local populations clustering fairly tightly around an axis representing their mean slope is more likely. This situation, illustrated in Figure 7c, will give rise to high interlocality correlations; and I suspect that it is the most common situation likely to be encountered. Undoubtedly, more complicated situations will occur. A single example is illustrated in Figure 7d. The slopes of the locality ellipses are identical, but the pattern of the means of the ellipses suggests a substantial correlation among the locality means yielding a different slope. Because such phenetic changes also involve translation of the phenetic surface away from its original principal axis, they are likely to be difficult to achieve, and thus will occur only rarely.

Patterns produced by populations without intralocality character correlations are shown in Figure 8. In the situation illustrated by Figure 8a the means of the intralocality distributions have been translated by evolutionary forces similar to those affecting Figure 7b. The pattern that the bivariate means have assumed is such that no correlation is shown among the localities. By contrast in Figure

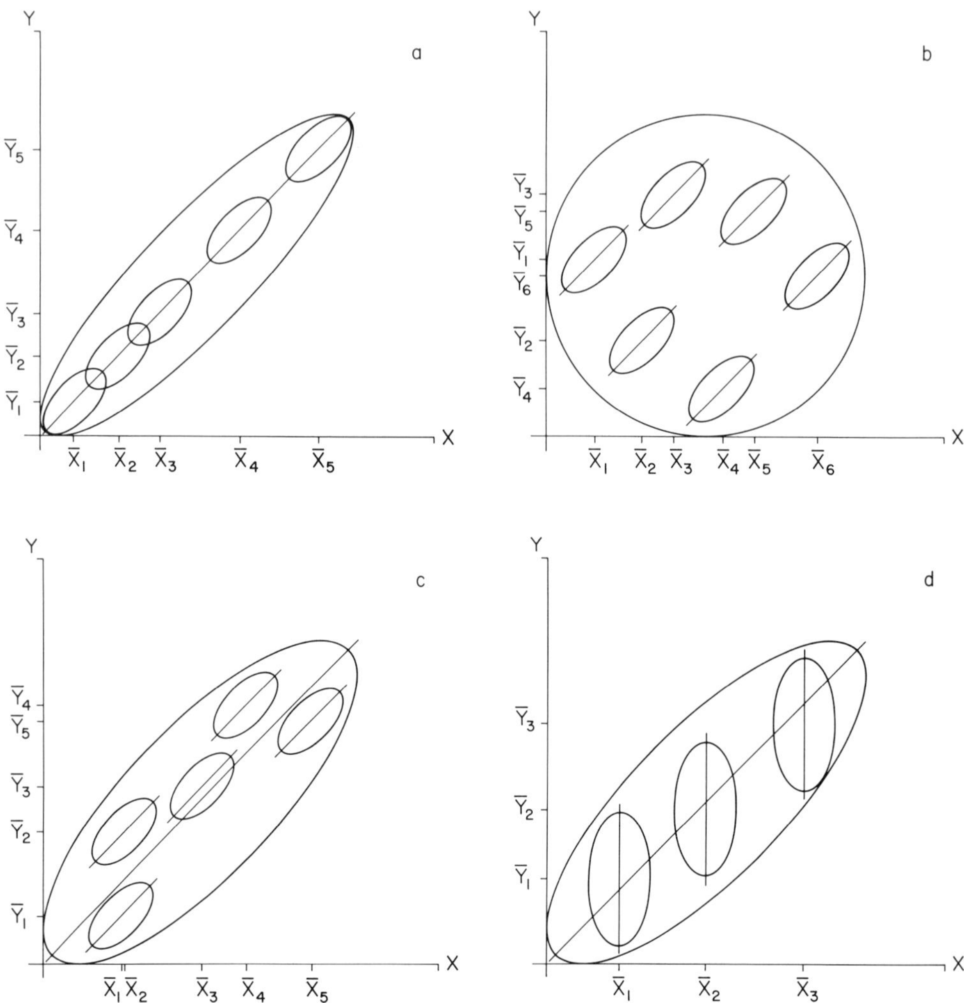

Fig. 7. Interlocality correlations and their relations to significant intralocality correlation. The axes define a phenetic space for characters X and Y. Each small ellipse is the phenetic distribution (isodensity contour) of a single population at one locality. Because their distributions are elliptical, the characters are correlated within localities. The large ellipse describes the correlation among localities; the large circle indicates absence of correlation. a. High interlocality and intralocality correlation. This is shown as the result of selection along the major axis of the within locality dispersion. b. No interlocality correlation coupled with high intralocality correlation. Selection has translated the original isodensity ellipse in various directions. Such a result should be difficult to obtain. c. Moderate interlocality correlation with high intralocality correlation. Some translation of original idodensity along minor axis, most translation along major axis. d. Correlation among locality means differing from intralocality correlations. One of many possible, but unlikely, combinations.

8b, the means are arranged in an elliptical fashion yielding correlation among localities. Selection shifting the bivariate means of uncorrelated characters in populations should be a simpler task than for correlated characters. Not only are there equal numbers of deviants in the four quadrants around the mean available for directional selection, but also the population need not undergo genetic rearrangements to break the correlated responses of the two characters in order to shift the distribution of phenotypes into a different portion of the phenetic surface. The populations represented by circles or ellipses in Figures 7b through 7d and Figure 8 have been shown as nonoverlapping to simplify the diagrams, but they are, of course, more likely to exhibit considerable phenetic overlap.

The notion that local populations move across phenetic space in response to different fitness surfaces does not mean that the position of locality means in phenetic space is arbitrary. It is unlikely that any locality can move beyond bounds set by the functional and developmental morphology of the organisms. Presumably, not all combinations of character means are equally acceptable. The adaptive topographies for the various localities are undoubtedly correlated.

The predictions concerning the relations between intra- and interpopulation correlation are borne out on two data sets which have been examined in this regard so far. Intralocality correlations of 16 morphological characters from a study of the geographic variation of the rabbit tick *Haemaphysalis leporispalustris* (Thomas, 1968) were plotted against interlocality component correlations of the same data (Figure 9). All high intralocality correlations give rise to high interlocality correlations, while the intermediate and low level intralocality correlations give rise to a wide range of interlocality correlations. These relationships are shown even more clearly in a study of the identical relations in the aphid *Pemphigus populitransversus* to be published elsewhere (Sokal and Riska, in preparation).

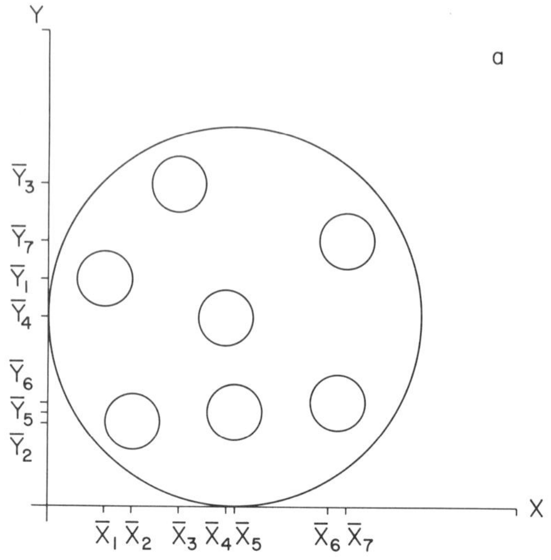

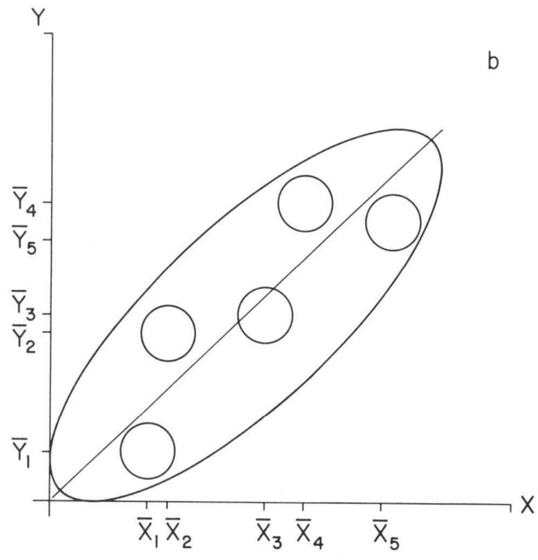

Fig. 8. Interlocality correlations when intralocality correlations are zero. The axes define a phenetic space for characters X and Y as in Figure 7. Each small circle is the phenetic distribution (isodensity contour) of a single population at one locality. Because their distributions are circular the characters are not correlated within localities. The large circle or ellipse describes the absence or presence of correlation among localities. a. No correlation among or within localities. The position of the population distributions in the phenetic space was effected by the translation of the original isodensity circle in various directions. Such a result should be easier to obtain than that of Figure 7b. b. Interlocality correlation with no intralocality correlation. This result should be as easy to obtain by selection as that of Figure 8a.

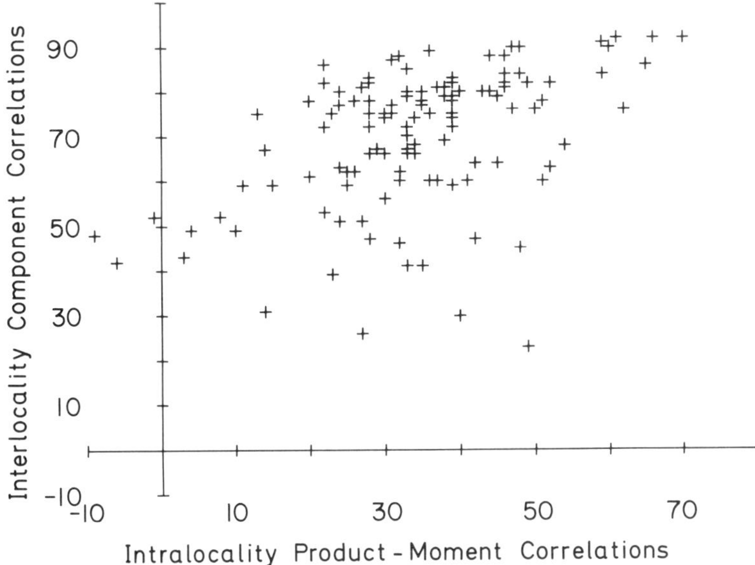

Fig. 9. Interlocality component correlations (ordinate) plotted against intralocality correlations (abscissa), from a geographic variation study of the rabbit tick *Haemaphysalis leporispalustris* (Thomas, 1968). Each point represents one pair from among the 16 morphological characters.

D. Conclusion

Let us return to our original question. I think we must at this stage bring in a Scottish verdict of "not proven" to the charge that the synthetic theory is insufficient to explain population differentiation. Personally, I am convinced that there are genomic rearrangements necessary for the manifested diversity to be realized. Charting the regularity of such rearrangements will require the combined efforts of developmental geneticists, morphometricians, and population geneticists in the years to come. I believe that not many years from now we shall be able to answer the question originally posed, and I suspect that the answer will be equivocal: more of the same leading to something new. There will be systems properties that are new, stemming from familiar phenomena that are more of the same.

Acknowledgements. This paper is dedicated to the memory of Professor Alfred E. Emerson, my teacher and a former President of the Society for the Study of Evolution.

I am indebted to Bruce Riska for discussing various points raised in this paper. He has taken little on faith and has forced me repeatedly to sharpen, or on occasion, abandon a given argument. Other aspects of this work benefited from discussions with Neal Oden and Dr. John Endler. Jacqueline Bird carried out some of the computations, Joyce Roe did the illustrations; and Barbara Scanlon turned snippets into typescript. I am grateful to them all.

This is Contribution No. 131 from the Graduate Program in Ecology and Evolution at the State University of New York at Stony Brook. Work on this paper was supported by Grant BMS 72-02211 A03 from the National Science Foundation.

References

Ayala, F.J., Tracey, M.L., Barr, L.G., McDonald, J.F., Pérez-Salas, S.: Genetic variation in natural populations of five *Drosophila* species and the hypothesis of the selective neutrality of protein polymorphisms. Genetics 77:343-384. (1974).

Berger, J.: Echinophilous ciliate protozoa as potential model systems in biogeography and population biology. Amer. Natur. (in press) (1978).

Cliff, A.D., Ord, J.K.: Spatial Autocorrelation. London: Pion, 1973, 175 pp.

Crow, J.K., Kimura, M.: An Introduction to Population Genetics Theory. New York: Harper and Row, 1970, 591 pp.

Dobzhansky, T.: Genetics of the Evolutionary Process. New York: Columbia Univ. Press, 1970, 505 pp.

Dobzhansky, T., Ayala, F.J., Stebbins, G.L., Valentine, J.W.: Evolution. San Francisco: W.H. Freeman, 1977, 572 pp.

Falconer, D.S.: Introduction to Quantitative Genetics. Edinburgh: Oliver and Boyd, 1960, 365 pp.

Frazzetta, T.H.: Complex Adaptations in Evolving Populations. Sunderland, Mass.: Sinauer Associates, 1975, 267 pp.

Friedlaender, J.S.: Patterns of Human Variation. Cambridge, Mass.: Harvard Univ. Press, 1975, 252 pp.

Johnson, M.S., Mickevich, M.F.: Variability and evolutionary rates of characters. Evolution 31:642-648. (1977).

Kerfoot, W.C.: The divergence of adjacent populations. Ecology 56: 1298-1313. (1975).

Kluge, A.G., Kerfoot, W.C.: The predictability and regularity of character divergence. Amer. Natur. 107:426-442. (1973).

Koehn, R.K., Eanes, W.F.: Molecular structure and protein variation within and among populations. Evolutionary Biology (in press). (1978).

Lewontin, R.C.: The Genetic Basis of Evolutionary Change. New York: Columbia Univ. Press, 1974, 346 pp.

Lynn, D.H., Berger, J.: Morphology, systematics and demic variation of *Pagiopyliella pacifica* Poljansky, 1951 (Ciliatea:Philasterina), an entocommensal of strongylocentroid echinoids. Trans. Amer. Microscop. Soc. 91:310-336. (1972).

Mayr, E.: Animal Species and Evolution. Cambridge, Mass.: Belknap, Harvard Univ. Press, 1963, 797 pp.

Mayr, E.: Populations, Species and Evolution. Cambridge, Mass.: Harvard Univ. Press, 1970, 453 pp.

Milkman, R.: The genetic basis of natural variation in *Drosophila melanogaster*. Adv. Genet. 15:55-114. (1970).

Rhoads, J.G., Friedlaender, J.S.: Language boundaries and biological differentiation on Bougainville: Multivariate analysis of variance. Proc. Nat. Acad. Sci. USA 72:2247-2250. (1975).

Riedl, R.: Die Ordnung des Lebendigen. Hamburg: Paul Parey, 1975, 372 pp.

Riedl, R.: A systems-analytical approach to macroevolutionary phenomena. Quart. Rev. Biol. 52:351-370. (1977).

Roughgarden, J.: Evolution of niche width. Amer. Natur. 106:683-718. (1972).

Royaltey, H.H., Astrachan, E., Sokal, R.R.: Tests for patterns in geographic variation. Geogr. Anal. 7:369-395. (1975).

Schaal, B.A.: Population structure and balancing selection in *Liatris cylindracea*. Ph.D. Dissertation. Yale University. (1974).

Schaal, B.A.: Population structure and local differentiation in *Liatris cylindracea*. Amer. Natur. 109:491-510. (1975).

Schaal, B.A., Levin, D.A.: The demographic genetics of *Liatris cylindracea* Michx. (Compositae). Amer. Natur. 110:191-206. (1976).

Selander, R.K., Kaufman, D.W.: Genetic structure of populations of the brown snail (*Helix aspersa*). I. Microgeographic variation. Evolution 29:385-401. (1975).

Senner, J.W.: Asymmetry and variability in populations of *Pemphigus populitransversus*. Ph.D. Dissertation. State University of New York at Stony Brook. (1977).

Sokal, R.R.: The Kluge-Kerfoot phenomenon reexamined. Amer. Natur. 110:1077-1091. (1976).

Sokal, R.R., Oden, N.L.: Spatial autocorrelation in biology: 1. Methodology. Biol. J. Linnean Soc. 10:199-228. (1978a).

Sokal, R.R., Oden, N.L.: Spatial autocorrelation in biology: 2. Some biological implications and four applications of evolutionary and ecological interest. Biol. J. Linnean Soc. 10: 229-249. (1978b).

Soulé, M.: Phenetics of natural populations. II. Asymmetry and evolution in a lizard. Amer. Natur. 101:141-160. (1967).

Soulé, M.: Phenetics of natural populations. III. Variation in insular populations of a lizard. Amer. Natur. 106:429-446. (1972).

Soulé, M., Baker, B.: Phenetics of natural populations. IV. The population asymmetry parameter in a butterfly *Coenonympha tullia*. Heredity 23:611-614. (1968).

Thomas, P.A.: Variation and covariation in characters of the rabbit tick, *Haemaphysalis leporispalustris*. Univ. Kansas Sci. Bull. 47:829-862. (1968).

Wright, S.: Evolution and the Genetics of Populations. II. The Theory of Gene Frequencies. Chicago: Univ. of Chicago Press, 1969, 511 pp.

Wright, S.: Evolution and the Genetics of Populations. III. Experimental Results and Evolutionary Deductions. Chicago: Univ. of Chicago Press, 1977, 613 pp.

Yablokov, A.V.: Variability of Mammals. New Delhi: Amerind Publishing Co., 1974, 350 pp.

Index